Lithium-Ion Supercapacitors

Electrochemical Energy Storage and Conversion

Series Editor:

Jiujun Zhang

Institute for Sustainable Energy/College of Sciences, Shanghai University, China

RECENTLY PUBLISHED TITLES

Electrochemical Polymer Electrolyte Membranes
Jianhua Fang, Jinli Qiao, David P. Wilkinson, and Jiujun Zhang

Electrochemical Supercapacitors for Energy Storage and Delivery: Fundamentals and Applications
Aiping Yu, Victor Chabot, and Jiujun Zhang

Photochemical Water Splitting: Materials and Applications
Neelu Chouhan, Ru-Shi Liu, and Jiujun Zhang

Metal–Air and Metal–Sulfur Batteries: Fundamentals and Applications
Vladimir Neburchilov and Jiujun Zhang

Electrochemical Reduction of Carbon Dioxide: Fundamentals and Technologies
Jinli Qiao, Yuyu Liu, and Jiujun Zhang

Electrolytes for Electrochemical Supercapacitors
Cheng Zhong, Yida Deng, Wenbin Hu, Daoming Sun, Xiaopeng Han, Jinli Qiao, and Jiujun Zhang

Solar Energy Conversion and Storage: Photochemical Modes
Suresh C. Ameta and Rakshit Ameta

Lead-Acid Battery Technologies: Fundamentals, Materials, and Applications
Joey Jung, Lei Zhang, and Jiujun Zhang

Lithium-Ion Batteries: Fundamentals and Applications
Yuping Wu

Graphene: Energy Storage and Conversion Applications
Zhaoping Liu and Xufeng Zhou

Proton Exchange Membrane Fuel Cells
Zhigang Qi

Lithium-Ion Supercapacitors
Fundamentals and Energy Applications

Edited by
Lei Zhang, David P. Wilkinson, Zhongwei Chen, and Jiujun Zhang

CRC Press is an imprint of the
Taylor & Francis Group, an **informa** business

CRC Press
Taylor & Francis Group
6000 Broken Sound Parkway NW, Suite 300
Boca Raton, FL 33487-2742

CRC Press is an imprint of Taylor & Francis Group, an Informa business

Printed and bound by CPI Group (UK) Ltd, Croydon, CR0 4YY

International Standard Book Number-13: 978-1-138-03219-4 (Hardback)

Library of Congress Cataloging-in-Publication Data

Names: Zhang, Lei (Chemist), author. | Wilkinson, David P., author. | Chen, Zhongwei (Professor), author. | Zhang, Jiujun, author.
Title: Lithium-ion supercapacitors : fundamentals and energy applications / [edited by] Lei Zhang, David P. Wilkinson, Zhongwei Chen and Jiujun Zhang.
Description: First edition. | Boca Raton : CRC Press/Taylor & Francis, [2018] | Series: Electrochemical energy storage and conversion | "A CRC title, part of the Taylor & Francis imprint, a member of the Taylor & Francis Group, the academic division of T&F Informa plc." | Includes bibliographical references and index.
Identifiers: LCCN 2018001390 | ISBN 9781138032194 (hardback) | ISBN 9781138032521 (ebook)
Subjects: LCSH: Lithium ion batteries. | Supercapacitors.
Classification: LCC TK2945.L58 L5996 2018 | DDC 621.31/2424--dc23
LC record available at https://lccn.loc.gov/2018001390

Visit the Taylor & Francis Web site at
http://www.taylorandfrancis.com

and the CRC Press Web site at
http://www.crcpress.com

Contents

Editors

Lei Zhang is a senior research officer at National Research Council Canada (NRC), a fellow of the Royal Society of Chemistry (FRSC), an adjunct professor of various universities, and a vice president of the International Academy of Electrochemical Energy Science (IAOEES). In 2004, she joined NRC Institute for Fuel Cell Innovation (NRC-IFCI) to help initiate the PEM Fuel Cell Program while she has carried out R&D of supercapacitors, metal-air batteries, Li-ion batteries, and hybrid batteries. She has co-authored more than 170 publications (>12,000 citations).

David P. Wilkinson is a professor and Canada Research Chair (Tier 1) in the Department of Chemical and Biological Engineering at the University of British Columbia (UBC), British Columbia, Canada. He has more than 80 issued patents and 170 refereed journal articles, and a number of edited books and book chapters, covering innovative research in these fields.

Zhongwei Chen is a professor and Canada Research Chair (Tier 1) in Advanced Materials for Clean Energy, a fellow of the Canadian Academy of Engineering, director of Collaborative Graduate Program in Nanotechnology, a director of Applied Nanomaterials & Clean Energy Laboratory at University of Waterloo, an associate editor of ACS Applied Materials & Interfaces, and the founder/associate chairman of the International Academy of Electrochemical Energy Science (IAOEES).

Jiujun Zhang is a professor and dean of the College of Sciences, and a dean of the Institute for Sustainable Energy at Shanghai University. He is a former principal research officer at the National Research Council Canada (NRC), a fellow member of the Academy of Science of the Royal Society of Canada (FRSC-CA), a fellow of the International Society of Electrochemistry (FISE), a fellow member of the Engineering Institute of Canada (FEIC), a fellow member of the Canadian Academy of Engineering (FCAE), a fellow of the Royal Society of Chemistry (FRSC-UK), and the founder/chairman of the International Academy of Electrochemical Energy Science (IAOEES).

Contributors

Muhammad Arif Khan
College of Science, School of Material Science and Engineering
Shanghai University
Shanghai, China

Kunfeng Chen
Changchun Institute of Applied Chemistry
Chinese Academy of Sciences
Changchun, China

Zhongwei Chen
Department of Chemical Engineering
Waterloo Institute for Nanotechnology, Waterloo Institute for Sustainable Energy
University of Waterloo
Waterloo, Ontario, Canada

Kun Feng
Department of Chemical Engineering
Waterloo Institute for Nanotechnology, Waterloo Institute for Sustainable Energy
University of Waterloo
Waterloo, Ontario, Canada

Zhenjiang He
School of Metallurgy and Environment
Central South University
Changsha, China

Xitong Liang
Changchun Institute of Applied Chemistry
Chinese Academy of Sciences
Changchun, China

Yuyu Liu
Institute of Sustainable Energy/ Department of Science
Shanghai University
Shanghai, China

Wei Pan
Changchun Institute of Applied Chemistry
Chinese Academy of Sciences
Changchun, China

Qinsi Shao
Sustainable Energy Research Institute
Shanghai University
Shanghai, China

Zhenyu Xing
Department of Chemical Engineering
Waterloo Institute for Nanotechnology, Waterloo Institute for Sustainable Energy
University of Waterloo
Waterloo, Ontario, Canada

Dongfeng Xue
Changchun Institute of Applied Chemistry
Chinese Academy of Sciences
Changchun, China

Wei Yan
Sustainable Energy Research Institute
Shanghai University
Shanghai, China

Dan Zhang
Institute of Sustainable Energy
Shanghai University
Shanghai, China

Jiujun Zhang
Institute for Sustainable Energy, College of Science
Shanghai University
Shanghai, China

Hongbin Zhao
College of Science, School of Chemistry
Shanghai University
Shanghai, China

Delun Zhu
Institute of Sustainable Energy
Shanghai University
Shanghai, China

1 Fundamentals of Lithium-Ion Supercapacitors

Zhenjiang He
Central South University

Yuyu Liu
Shanghai University

CONTENTS

1.1 HISTORY

In the nineteenth century, German scientist Helmholtz discovered the double electric layer that forms on the interface between a charged electrode and an electrolyte solution. In the middle of the twentieth century, Grahame further improved the double electrode layer theory and laid the foundation for the application of supercapacitors. Since the 1990s, with the development and popularity of electric vehicles, development of high-energy pulsed-power sources are receiving more attention. At present, traditional capacitors have low energy density, and lithium-ion batteries have low power density [1–4]. Therefore, a single application of traditional capacitors or lithium-ion batteries can hardly meet these demands of specific electric power tools. To solve this problem, two kinds of methods have been put forward. First, the combined utilization of traditional capacitors and lithium-ion batteries has been proposed to satisfy these demands. Combining the high power density of a traditional capacitor with the high energy density of a lithium-ion battery can satisfy these demands to a certain extent. However, this combination requires other additional accessories to meet regulatory functions, which reduces the overall energy density by increasing the weight of the device. Second, new types of hybrid capacitors based on capacitor technology and the electrochemical principle have been developed to satisfy these demands [5–7].

Since the beginning of the 1990s, many famous research institutions and large corporations moved their research focus from electric double-layer capacitors (EDLCs) to new style capacitors. In 1990, Giner, Inc. reported an aqueous pseudo-capacitor, which used noble metal oxides as electrode materials [8]. In order to further improve the specific capacity of capacitors, D.A. Evans, in 1995, proposed a significant concept of electrochemical hybrid capacitor combining ideal polarized electrodes and Faraday electrodes [9]. In 1997, a Russian company called ESMA publicized a new hybrid capacitor system (NiOOH/activated carbon [AC]), which revealed a novel technology that integrates battery materials and capacitive materials for electrochemical devices. In 2001, G.G. Amamcci reported a nonaqueous hybrid capacitor that used the lithium-ion battery material ($Li_4Ti_5O_{12}$) and AC as electrode materials, which is regarded as a milestone in the development of electrochemical hybrid capacitor. In 2005, Fuji Heavy Industries (FHI) publicized a novel electrochemical hybrid capacitor, which added lithium-ions to improve energy density that they named the lithium-ion supercapacitor (LISC). The key point of the LISC developed by FHI is doping anode materials (polyacene) with lithium in advance, which resulted in improved energy density of the anode by more than 30 times compared to the AC. In addition, the lithium doping can dramatically decrease the anode potential, which makes the individual cell voltage increase about 1.5 times, further improving the energy density. In 2006, Hatozaki reported a new style LISC, which used AC as the cathode material and a carbon material preintercalated with lithium as the anode material, with the operating voltage of this LISC reaching 3.8 V in organic electrolytes.

In the twenty-first century, increasing attention has been paid to the research of LISC. Nowadays, carbon materials (such as graphene, AC, and graphite), transition metal oxide, and transition metal sulfide have been widely studied and developed as electrode active materials because of the excellent electrochemical performance they display [10–15]. Many methods (such as hydrothermal treating, heat treatment, and atmosphere control) have been applied to design the morphology and structure of electrode materials and to enhance their electrochemical properties. Against the background of conventional energy crisis and environmental damage, the research on LISC becomes a popular area in the twenty-first century, and most of the countries in the world have invested many human and financial resources on it.

1.2 LITHIUM-ION SUPERCAPACITOR PHYSICS/CHEMISTRY/ELECTROCHEMISTRY

The LISC consists of an anode, cathode, electrolyte, separator, current collector, capacitor pack, etc. A LISC, in essence, is a hybrid capacitor, which can be understood as the active material of one or two electrodes of a supercapacitor substituted by a lithium-ion battery material and operated in an electrolyte with the lithium content. Obviously, as a hybrid capacitor, the energy storage mechanism of LISCs includes not only redox reaction of lithium-ion batteries, like lithium intercalation/transformation, alloying, etc. but also electric double layer and faradaic pseudocapacitance.

The energy storage of EDLCs is dependent on the electrostatic adsorption of cations and anions on the surface of an electrode. During the charge process, cations and anions will migrate to the cathode or anode under the electric field, separately. In turn, during the discharge process, these ions will be desorbed from cathode or anode and migrate the opposite way. The reaction process of double layer capacitor can be displayed as follows:

$$E_A + A^- \leftrightharpoons E_A^+/A^- + e^- \tag{1.1}$$

$$E_C + C^+ \leftrightharpoons E_C^-/C^+ + e^-, \tag{1.2}$$

where E_A and E_C are the anode and cathode, respectively; A^- and C^+, the anion and cation; and e^-, the electron.

In general, the activated carbons with high specific surface areas are utilized as the electrode material of electric double layer capacitors, and the capacitance of one electrode is calculated using the following equation:

$$C = \frac{A\left(\varepsilon_r \varepsilon_0\right)}{d}, \tag{1.3}$$

where C is the capacitance of one electrode; A, the effective specific surface area of activated carbon; ε_r, the permittivity of electrolyte solution; ε_0, the permittivity of vacuum; and d, the distance of electrodes.

The basis of faradaic pseudocapacitance is the redox reaction between electrode materials and ions in the electrolyte, which can be further subdivided into three different types: redox reaction of transition metal oxides, protonation reaction of conducting polymers, and reversible absorption of hydrogen ions. Take MnO_2 as an example; the reaction process of faradaic pseudocapacitance can be displayed as follows:

$$MnO_2 + M^+ + e^- \leftrightharpoons MnOOH, \tag{1.4}$$

where M^+ = Li, Na, K.

Until now, there are many mechanisms of energy storage in a lithium-ion battery, such as intercalation reactions, alloying reactions, phase transformations, conversion reactions, free radical reactions, electrodeposition, interfacial interactions, and surface adsorption.

The typical principle of different capacitors is illustrated in Figure 1.1 [16]. The configuration of LISCs can be divided into two categories based on the electrolyte medium (aqueous and nonaqueous). Meanwhile, the LISC can be divided into two types, symmetric and asymmetric systems, based on the combination of capacitor materials and lithium-ion battery materials.

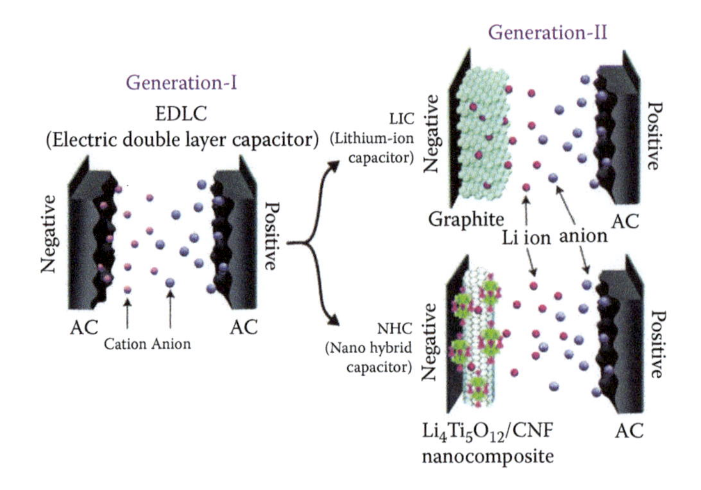

FIGURE 1.1 The typical schematic diagram of electric double-layer capacitor (EDLC), lithium-ion capacitor (LIC), and lithium-ion based hybrid capacitor (Li-HEC, which is the same as a lithium-ion supercapacitor [LISC]) [16] (Reprinted from Aravindan, V. et al., *Chemical Reviews*, 2014, 114, 11619 with permission).

1.3 LITHIUM-ION SUPERCAPACITOR COMPONENTS (ANODE, CATHODE, ELECTROLYTE, SEPARATOR/ MEMBRANES, CURRENT COLLECTOR)

LISCs are a class of advanced energy storage devices combining second batteries and EDLCs. Usually, EDLCs utilize high surface area AC as both cathode and anode material. LISCs utilize AC as the anode material and an inert compound, which supports the fast reversible intercalation of lithium-ions, as a cathode material. Both of these anode and cathode materials are mixed with the binder (and conductive materials) to form a slurry and then coated on the Cu foil, Al foil, Ti foil, or stainless steel substrate [17–19]. In LISCs, a membrane is utilized to separate the positive and negative electrodes and to prevent electron conduction and short circuits. However, the isolation property of the separator means that the small pore diameter will resist the movement of electrolytes. Therefore, the relationship of pore diameter and electrochemical properties is the main research topic [20,21]. The membrane should meet the following requirements, taking into account the processability and usability: (a) separator papers should be insulated and isolated in order to prevent short circuits while also letting the electrolyte pass through easily; (b) uniform thickness and homogenous pore-size distribution; (c) chemical stability in electrolyte, dimensional stability, certain level of mechanical strength, and thermostability; (d) good wettability and electrolyte storage function; and (e) the resistance should be as small as possible which means perfect ion transportation. Nowadays, the most common separator paper materials are polypropylene and cellulose. Based on how it is structured, separator paper can be divided into single layer structure and hybrid structure (two-layer and three-layer). The two general processes that are used to produce separator papers are dry-laid nonwoven fabric and wet-laid nonwoven fabric.

AC is one of the most popular capacitive materials. According to the working potential, AC can serve as not only a cathode material but also as an anode material. The specific capacity of AC has important effects on the capacitance of LISCs. Recently, the main method to improve the capacitance of AC is by modifying the open framework structure and surface structure. The charge storage capability of AC is dependent on the electrostatic adsorption on the surface of AC. However, continuously increasing the specific surface area of AC cannot result in an infinite increase in the charge storage as the pore diameter becomes too small for proper electrolyte infiltration. Furthermore, the increase of specific surface area and porosity will result in the decrease of skeletal density and then deteriorate the electric conductivity. Therefore, how to design and control the specific surface area and porosity is the key to improving the properties of AC.

The typical intercalation compound applied in LISC can be divided into four categories: (1) lithium-ion containing metal oxides, (2) polyanionic compounds, (3) graphite, and (4) transition metal oxides. Generally, lithium-ion containing metal oxides and polyanionic compounds serve as cathode materials; on the contrary, graphite and transition metal oxides are used as anode materials. The application of these four kinds of materials in LISC will be introduced in the following sections.

1.3.1 Lithium-Ion Containing Metal Oxides

According to crystal structure, lithium-ion containing metal oxides $LiMO_x$ (M = Mn, Co, Ni, etc.) can be divided into two categories: layered oxides ($LiMO_2$) and spinel oxides (LiM_2O_4). Lithium cobalt oxide ($LiCoO_2$) is the most widely used commercial material, and its theoretical capacity is around 274 mAh g^{-1}. However, the practical capacity of $LiCoO_2$ is about half of the theoretical value due to its structural instability at complete lithium delithiation [22,23]. According to the literature, element doping using Zr, Mg, Mo, Sr, V, and Al or surface modification with $LiMn_2O_4$, MgO, $AlPO_4$, TiO_2, Al_2O_3, SiO_2, B_2O_3, and LiPON polyimide is an efficient way to raise the specific capacity of $LiCoO_2$ [23–31]. However, the high price and toxicity of cobalt have limited further application of $LiCoO_2$. Recently, many compounds have been developed as cathode materials to replace the commercial $LiCoO_2$, such as $LiNi_{1/3}Co_{1/3}Mn_{1/3}O_2$, and $LiNi_{0.5}Mn_{1.5}O_4$ [32–38].

1.3.2 Polyanionic Compounds

$Li_xM_y(ZO_4)$ (M is transition metal, Z = P, S, Si, Mo, W, etc.) has been thought of as the most promising cathode material in the field of power batteries. In 1997, K.S. Goodenough and his colleague found the olivine-type phosphate compound $LiFePO_4$ (about 170 mAh g^{-1}), which attracted extensive attention due to low cost, abundant raw materials, low toxicity, long cycle life, and safety [39]. Unfortunately, $LiFePO_4$ exhibits a low-intrinsic electronic conductivity, about 10^{-9}–10^{-10} S/cm, which is much lower than that of $LiCoO_2$ (10^{-3} S/cm) and is the main obstacle to its application. Nowadays, the main methods adopted to circumvent this drawback include conductive agent coating and minimizing the particle size of $LiFePO_4$, etc. [39–41]. Conducive agent coating, such as carbon materials, can dramatically enhance the electronic conductivity and resist the corrosion on the surface of $LiFePO_4$ particles. Reduced particle sizes are obtained through the synthesis of nanoparticles by different methods, and can facilitate the intercalation/deintercalation of lithium-ions and enhance its ionic conductivity.

1.3.3 Graphite

Graphite is the most popular anode material applied in lithium-ion batteries. However, the specific surface area of graphite is much smaller than that of porous carbon, which result in the low rate capacity of graphite. S.R. Sivakkumar reported that the rate capacity can be enhanced by increasing the specific surface area of commercial graphite from 9.4 to 416 $m^2\ g^{-1}$ using a ball mill. Recently, petroleum coke, carbon nanotubes, graphene, etc. have been proposed as a substitute for commercial graphite [42,43]. J.J. Ren compared the electrochemical performances of pre-lithiated graphene nanosheets and conventional graphite as negative electrode materials for LISC, and found pre-lithiated graphene nanosheets exhibit excellent specific capacitance, cycle stability, and rate capability [42].

1.3.4 Transition Metal Oxides

Recently, $Li_4Ti_5O_{12}$, a so-called zero-strain lithium insertion oxide with a theoretical capacity of 175 mAh g^{-1}, has acquired considerable attention as an anode material of LISC. The Li insertion potential is around 1.55 V for $Li_4Ti_5O_{12}$, which is well above the formation potential of metallic lithium and can effectively suppress the decomposition of electrolyte. As a result, $Li_4Ti_5O_{12}$ displays stable discharge potential, excellent cycle stability, and is especially safe to use. Unfortunately, the poor electronic conductivity of $Li_4Ti_5O_{12}$ anode results in its poor rate performance. Conducive agent coating and minimizing the particle size of $Li_4Ti_5O_{12}$ particles are effective ways to enhance its electronic conductivity and then improve the rate performance. H.W. Wang created a freestanding TiO_2 nanobelt array on Ti foil, which can enhance the transportation of electron and ion and then exhibit high-rate capability and stable cycle performance [19]. H.L. Wang synthesized a monolayer of MnO nanocrystallites mechanically anchored by pore-surface terminations of 3D arrays of graphene-like carbon nanosheets, which achieved high energy, high rate, and high capacity characteristics [6]. Besides, many other transition metal oxides, such as Bi_2O_3, Fe_3O_4, MoO_2, and V_2O_5, have been explored as anode materials [4,12,15,17,18].

Generally, LISCs can work in not only aqueous but also in a nonaqueous mediums. Aqueous electrolytes display low viscosity and high ion conductivity. However, the work potential of electrode materials was restricted to 1 V, and thus the energy density of electrode materials could not be brought into full play. X_2SO_4, XOH, and XNO_3 (X = Li, Na, K, etc.) have frequently been used as electrolytes. On the contrary, the voltage window can reach to 3 V, or even 4 V, with nonaqueous electrolytes (usually dissolved $LiPF_6$ in ethylene carbonate and dimethyl carbonate), which can dramatically improve the energy storage of LISC. But the high viscosity of nonaqueous electrolyte results in the low ion conductivity, affecting the power density.

1.4 ELECTROCHEMICAL PERFORMANCE (ENERGY/POWER DENSITIES, CAPACITY, CHARGE–DISCHARGE, CYCLE LIFE, FAILURE MODE ANALYSIS, MODELING, CODES, AND STANDARDS)

Energy and power densities are the main characteristics of LISC application, which can be calculated using the following formulas:

$$E = \frac{C_{cell}\ V^2}{2 \times 3.6} \text{ and} \tag{1.5}$$

$$P = \frac{3600 \times E}{\Delta t}, \tag{1.6}$$

where E is the energy density of LISC, W h kg^{-1}; C_{cell}, the specific capacitance of the LISC, F g^{-1}; V, the potential window of discharge current, V; P, the power density of LISC, W kg^{-1}; and Δt, the discharging time, s.

Generally, cyclic voltammetry and chronopotentiometry are utilized to measure the specific capacitance.

1.4.1 Cyclic Voltammetry

Cyclic voltammetry is an essential research method of electrochemistry that has been utilized to study the mechanism of electrode reactions and dynamic factors of the electrode processes. The electrochemical reaction in the electrode and its reversibility can be evaluated by the location and peak current of redox peaks. According to the cyclic voltammetry curve, the specific capacitance can be calculated using the following formula:

$$C_s = \frac{1}{m \times v \times (\varphi_2 - \varphi_1)} \int_{\varphi_1}^{\varphi_2} I \mathrm{d}V, \tag{1.7}$$

where C_s is the specific capacitance of the electrode material, F g^{-1}; m, the mass of total active material mass of electrodes, g; I, the constant charge or discharging current, A; v, the potential scan rate, mV s^{-1}; dV, the potential drop at a constant discharge current, V; φ_1, the initial potential, V; and φ_2, the final potential, V.

1.4.2 Chronopotentiometry

The fundamental principle of chronopotentiometry is that the electrode is charged and discharged under a constant current, so that a potential–time curve (GC curve) is obtained. The geometric shape of ideal GC curves is a symmetrical isosceles triangle. However, there are some differences between the actual and ideal GC curves, which is due to the effect of diffusion and pseudocapacitance. Under a three-electrode system, the specific capacitance of the electrode can be calculated using the following formula:

$$C_3 = \frac{I \times \Delta t}{m \times \Delta V}. \tag{1.8}$$

Under a double-electrode system, the specific capacitance of the electrode can be calculated using the following formula:

$$C_2 = \frac{4 \times I \times \Delta t}{m \times \Delta V}. \tag{1.9}$$

The specific capacitance of capacitor can be calculated using the following formula:

$$C_{cell} = \frac{1}{4} C_2, \tag{1.10}$$

where C_3 and C_2 are the specific capacitances of the electrode material under the three-electrode and double-electrode system, respectively; C_{cell}, the specific capacitance of the capacitor, F g^{-1}; m, the mass of total active material mass of electrodes, g; I, the constant charge or discharging current, A; Δt, the discharging time, s; ΔV, the potential window, V.

Nowadays, increased energy and power density have been reported, and factors, such as types of electrode materials, pre-lithiation degree, and work potential window, that can affect the electrochemical properties of LISCs have been investigated [11,13,44–52]. Amatucci et al. investigated the electrochemical properties of nanostructured Li_4TiO_5 as a negative electrode material and AC as a positive electrode material. It was found that the reaction rate was improved after the application of high rate nanostructured Li_4TiO_5, while retaining 90% capacity at 10 C charge rates and 85%–90% capacity retention after 5,000 cycles. The energy density of 20 W h kg^{-1} was calculated under a flat plate cell configuration [53]. Han et al. modified graphene with MoO_2 (G-MoO_2), which resulted in a material that displays high specific capacitance, good rate capability, and excellent cycle properties. The specific capacitance of G-MoO_2 maintained 91.2% capacity after 500 cycles at a current density of 1,000 mA exhibits an energy density of 33.2 W h kg^{-1} at a power density of 3,000 W kg^{-1} [18]. Cai et al. reported an anode TiO_2 with modified morphology and structure that exhibits a high energy density of 80 W h kg^{-1} [54]. A LISC assembled with graphite and AC delivers an energy density of 103 W h kg^{-1}, which was reported by Khomenko et al. [55]. According to these reports, the capacity fading during continuous cycles mainly occurs in the initial cycles, which is proposed to be caused by the formation of solid electrolyte interphase film on anode surface [11,48,56,57]. One approach to improve the energy density is by increasing the electrode potential window. However, under high potential, the crystal structure of electrode materials collapses easily, and the electrode will become unstable. On the other side, low potential could accelerate the formation of solid electrolyte interphase. What is worse, dendritic lithium may grow crystalline defects, such as kinks, grain boundaries, and interfaces, which could penetrate the separator membrane and cause short circuiting risks [58,59]. Therefore, the modification of electrochemical performance of LISCs is a comprehensive work that is concerned with optimizing the potential window, designing morphology and structure of electrode materials, even developing advanced materials, new capacitor systems, etc.

1.5 SUMMARY

The LISC is an ideal choice for electric vehicles in the future. It has several times higher energy density than that of an electrical double-layer capacitors and has a better rate capacity than that of a lithium-ion batteries. The technology of LISC is still in the research phase, but it has many promising applications. However, a LISC possesses a distinguished high energy and power density, and they have strict requirements with electrode materials. Therefore, the current research priority on LISCs is to develop composite electrode materials, which exhibit high specific capacitance, high potential window, high power density, long cycle life, etc. Obviously, integration of supercapacitors and second batteries becomes an inevitable trend of developing technology in the future. Someday, the development of supercapacitors and second batteries may unite.

REFERENCES

1. X. Zuo, J. Zhu, P. Müller-Buschbaum, Y.-J. Cheng, *Nano Energy*, 31 (2017) 113–143.
2. D. Chen, F. Zheng, L. Li, M. Chen, X. Zhong, W. Li, L. Lu, *Journal of Power Sources*, 341 (2017) 147–155.
3. D.-W. Wang, H.-T. Fang, F. Li, Z.-G. Chen, Q.-S. Zhong, G.Q. Lu, H.-M. Cheng, *Advanced Functional Materials*, 18 (2008) 3787–3793.
4. L. Ding, Q. Xin, X. Dai, J. Zhang, J. Qiao, *Ionics*, 19 (2013) 1415–1422.
5. T. Aida, K. Yamada, M. Morita, *Electrochemical and Solid-State Letters*, 9 (2006) A534.
6. H. Wang, Z. Xu, Z. Li, K. Cui, J. Ding, A. Kohandehghan, X. Tan, B. Zahiri, B.C. Olsen, C.M.B. Holt, D. Mitlin, *Nano Letters*, 14 (2014) 1987–1994.
7. J. Ding, H. Wang, Z. Li, K. Cui, D. Karpuzov, X. Tan, A. Kohandehghan, D. Mitlin, *Energy & Environmental Science*, 8 (2015) 941–955.
8. S. Sarangapani, P. Lessner, J. Forchione, A. Griffith, A.B. Laconti, *Journal of Power Sources*, 29 (1990) 355–364.
9. J.P. Zheng, T.R. Jow, *Journal of Power Sources*, 62 (1996) 155–159.
10. J. Zhang, Z. Shi, C. Wang, *Electrochimica Acta*, 125 (2014) 22–28.
11. X. Sun, X. Zhang, W. Liu, K. Wang, C. Li, Z. Li, Y. Ma, *Electrochimica Acta*, 235 (2017) 158–166.
12. V. Aravindan, Y.L. Cheah, W.F. Mak, G. Wee, B.V.R. Chowdari, S. Madhavi, *ChemPlusChem*, 00 (2012) 1–7.
13. M. Schroeder, M. Winter, S. Passerini, A. Balducci, *Journal of Power Sources*, 238 (2013) 388–394.
14. S. Sivakkumar, A. Pandolfo, *Electrochimica Acta*, 65 (2012) 280–287.
15. D. Qu, L. Wang, D. Zheng, L. Xiao, B. Deng, D. Qu, *Journal of Power Sources*, 269 (2014) 129–135.
16. V. Aravindan, J. Gnanaraj, Y.S. Lee, S. Madhavi, *Chemical Reviews*, 114 (2014) 11619–11635.
17. Y.L. Cheah, R.V. Hagen, V. Aravindan, R. Fiz, S. Mathur, S. Madhavi, *Nano Energy*, 2 (2013) 57–64.

18. P. Han, W. Ma, S. Pang, Q. Kong, J. Yao, C. Bi, G. Cui, *Journal of Materials Chemistry A*, 1 (2013) 5949–5954.
19. H. Wang, C. Guan, X. Wang, H.J. Fan, *Small*, 15 (2015) 1470–1477.
20. M. Winter, R.J. Brodd, *Chemical Reviews*, 104 (2004) 4245–4269.
21. B.E. Conway, *Journal of the Electrochemical Society*, 138 (1991) 1539–1548.
22. Y. Furushima, C. Yanagisawa, T. Nakagawa, Y. Aoki, N. Muraki, *Journal of Power Sources*, 196 (2011) 2260–2263.
23. S.A. Needham, G.X. Wang, H.K. Liu, V.A. Drozd, R.S. Liu, *Journal of Power Sources*, 174 (2007) 828–831.
24. S. Valanarasu, R. Chandramohan, *Journal of Alloys and Compounds*, 494 (2010) 434–438.
25. J. Cho, G. Kim, *Electrochemical and Solid-State Letters*, 2(6) (1999) 253–255.
26. M. Mladenov, R. Stoyanova, E. Zhecheva, S. Vassilev, *Electrochemistry Communications*, 3 (2001) 410–416.
27. M.G. Kim, C.H. Yo, *Journal of Physical Chemistry B*, 103 (1999) 6457–6465.
28. Y.J. Kim, J. Cho, T.-J. Kim, B. Park, *Journal of the Electrochemical Society*, 150 (2003) A1723.
29. K.-H. Choi, J.-H. Jeon, H.-K. Park, S.-M. Lee, *Journal of Power Sources*, 195 (2010) 8317–8321.
30. J.-H. Park, J.-S. Kim, E.-G. Shim, K.-W. Park, Y.T. Hong, Y.-S. Lee, S.-Y. Lee, *Electrochemistry Communications*, 12 (2010) 1099–1102.
31. J. Cho, *Journal of Power Sources*, 126 (2004) 186–189.
32. Z. Lu, Z. Chen, J.R. Dahn, *Chemistry of Materials*, 15 (2003) 3214–3220.
33. W. Zhao, L. Xiong, Y. Xu, H. Li, Z. Ren, *Journal of Power Sources*, 349 (2017) 11–17.
34. Y. Ma, Y. Zhou, C. Du, P. Zuo, X. Cheng, L. Han, D. Nordlund, Y. Gao, G. Yin, H.L. Xin, M.M. Doeff, F. Lin, G. Chen, *Chemistry of Materials*, 29 (2017) 2141–2149.
35. G.B. Zhong, Y.Y. Wang, Y.Q. Yu, C.H. Chen, *Journal of Power Sources*, 205 (2012) 385–393.
36. S. Yang, X. Wang, X. Yang, Y. Bai, Z. Liu, H. Shu, Q. Wei, *Electrochimica Acta*, 66 (2012) 88–93.
37. L.-J. Li, Z.-X. Wang, Q.-C. Liu, C. Ye, Z.-Y. Chen, L. Gong, *Electrochimica Acta*, 77 (2012) 89–96.
38. Z. He, Z. Wang, H. Chen, Z. Huang, X. Li, H. Guo, R. Wang, *Journal of Power Sources*, 299 (2015) 334–341.
39. H.-H. Chang, C.-C. Chang, C.-Y. Su, H.-C. Wu, M.-H. Yang, N.-L. Wu, *Journal of Power Sources*, 185 (2008) 466–472.
40. Y. Cui, X. Zhao, R. Guo, *Journal of Alloys and Compounds*, 490 (2010) 236–240.
41. X.-M. Guan, G.J. Li, C.-Y. Li, R.-M. Ren, *Transactions of Nonferrous Metals Society of China*, 27 (2017) 141–147.
42. J. Ren, L. Su, X. Qin, M. Yang, J. Wei, Z. Zhou, P. Shen, *Journal of Power Sources*, 264 (2014) 108–113.
43. Y. Ma, H. Chang, M. Zhang, Y. Chen, *Advanced Materials*, 27 (2015) 5296–5308.
44. T. Aida, I. Murayama, K. Yamada, M. Morita, *Journal of the Electrochemical Society*, 154 (2007) A798.
45. M. Schroeder, S. Menne, J. Ségalini, D. Saurel, M. Casas-Cabanas, S. Passerini, M. Winter, A. Balducci, *Journal of Power Sources*, 266 (2014) 250–258.
46. W.J. Cao, J. Shih, J.P. Zheng, T. Doung, *Journal of Power Sources*, 257 (2014) 388–393.
47. J. Zhang, X. Liu, J. Wang, J. Shi, Z. Shi, *Electrochimica Acta*, 187 (2016) 134–142.
48. J.-H. Kim, J.-S. Kim, Y.-G. Lim, J.-G. Lee, Y.-J. Kim, *Journal of Power Sources*, 196 (2011) 10490–10495.
49. W.J. Cao, J.P. Zheng, *Journal of the Electrochemical Society*, 160 (2013) A1572–A1576.
50. S.R. Sivakkumar, A.G. Pandolfo, *Electrochimica Acta*, 65 (2012) 280–287.

51. W.J. Cao, J.P. Zheng, *Journal of Power Sources*, 213 (2012) 180–185.
52. W. Cao, Y. Li, B. Fitch, J. Shih, T. Doung, J. Zheng, *Journal of Power Sources*, 268 (2014) 841–847.
53. G.G. Amatucci, F. Badway, A. Du Pasquier, T. Zheng, *Journal of the Electrochemical Society*, 148 (2001) A930.
54. Y. Cai, B. Zhao, J. Wang, Z. Shao, *Journal of Power Sources*, 253 (2014) 80–89.
55. V. Khomenko, E. Raymundo-Piñero, F. Béguin, *Journal of Power Sources*, 177 (2008) 643–651.
56. Z. Shi, Z. Jin, W. Jing, J. Shi, C. Wang, *Electrochimica Acta*, 153 (2015) 476–483.
57. F. Béguin, F. Chevallier, C. Vix, S. Saadallah, J.N. Rouzaud, E. Frackowiak, *Journal of Physics and Chemistry of Solids*, 65 (2004) 211–217.
58. M. Ouyang, Z. Chu, L. Lu, J. Li, X. Han, X. Feng, G. Liu, *Journal of Power Sources*, 286 (2015) 309–320.
59. J. Steiger, D. Kramer, R. Mönig, *Journal of Power Sources*, 261 (2014) 112–119.

2 Anodes of Lithium-Ion Supercapacitors

Wei Yan and Qinsi Shao
Shanghai University

CONTENTS

The technological breakthrough of anode materials for lithium secondary battery in the end of the 1980s and early 1990s resulted in the birth and commercialization of lithium-ion battery (LIB) [1]. However, despite the high energy densities, the power densities and cycling lifetimes of LIBs are limited due to the sluggish solid-state lithium-ion transport. Although supercapacitors (electrical double-layer capacitors [EDLCs]) can provide powerful bursts and good durability, their energy storage is low because of the limited specific surface area of the electrode materials. Lithium-ion supercapacitors (LISCs), consisting of an LIB electrode and an EDLC electrode in a lithium salt containing organic electrolyte, are expected to bridge the gap between LIBs and EDLCs and become the ultimate power source for hybrid vehicles (HEVs) and electrical vehicles (EVs) [2]. Due to the wide working voltage window of the organic electrolytes, LISCs can store more energy than EDLCs. And because of the introduction of capacitor-type electrodes, LISCs can achieve greater power densities than LIBs. Activated carbon (AC) is the unanimous choice for the

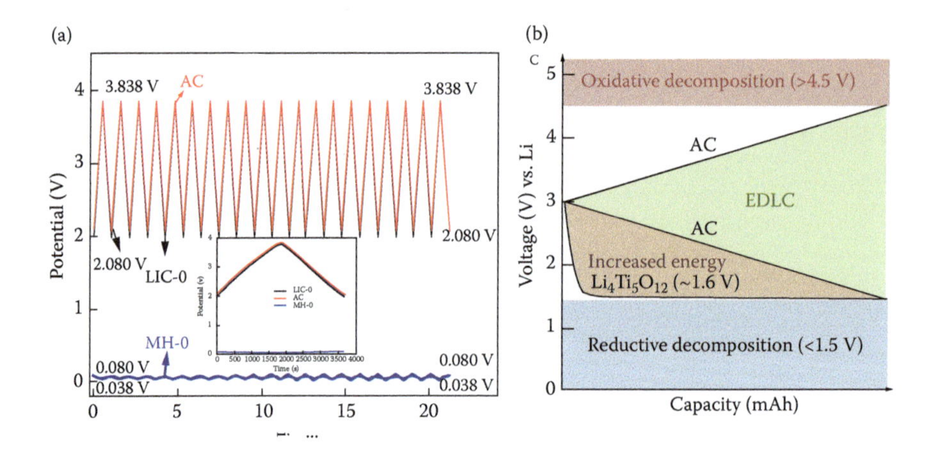

FIGURE 2.1 (a) The charge/discharge profiles of the carbon anode, AC cathode and the carbon/AC LISC [3]; (b) typical electrochemical profiles of anode and cathode in AC/AC EDLC and LTO/AC LISC [4].

capacitor-type electrode material owing to its high surface area, good electronic conductivity, eco-friendliness, and low cost; a lot of research has been or is being carried out on the battery-type anode electrode materials (Figure 2.1)

2.1 CARBON

A battery-type anode electrode material should satisfy the requirements such as low working voltage, high specific capacity, excellent rate capability, and long cycle life. Among all the battery-type anode materials that have been studied, carbon is still the dominant commercially available anode material. The most prominent feature of the LISCs based on carbon battery-type anodes is the wide working voltage. As the lithiated carbons have the potential close to that of metallic lithium, the carbon-anode LISCs usually work between ~2 and 4 V; meanwhile, the AC cathodes are cycled between 2 and 4 V. The full use of the AC electrodes can improve the electrochemical performance of the LISCs. On the contrary, the LISCs based on transition metal oxide anodes such as TiO_2, can only be charged/discharged between 1.2 and 3.2 V because the intercalation/deintercalation of lithium-ions in $Li_4Ti_5O_{12}$ (LTO) takes place at ~1.6 V. In such conditions, The AC cathodes are usually cycled between ~3 and 4 V, resulting in the under usage of the AC electrodes [5]. Nowadays, carbons such as graphite, hard carbon (HC), and graphene have been used for the battery-type anode materials in LISCs.

2.1.1 Graphite

The mechanism of lithium intercalation into graphite is well known [6,7], and it proceeds through several well-identified stages which can be characterized by X-ray diffraction (XRD) measurements and potential plateaus on open-circuit voltage (OCV) curves. Ohzuku assumed the existence of first, second, third, fourth, and eighth stages and the transition of each stage [7,8]. The theoretical capacity of natural flake graphite is 372 mAh/g (based on the formation of LiC_6, first stage). The reversible capacity is only about 300–350 mAh/g because of the impurities and defects in graphite powders (Figure 2.2).

Khomenko et al. constructed a LISC using commercial graphite and AC as the negative and positive electrodes, respectively [5]. The electrochemical properties of two types of commercial graphite (potato shape graphite and prismatic shape graphite) were evaluated. As lithium-ion intercalation occurs mainly at the edge sites of graphite, 3D structure (potato shape) could intercalate more lithium-ions than 2D structure (prismatic shape) at high current densities [10]. The LISC was assembled with the potato shape graphite anode and AC cathode. The mass of the electrodes was balanced to achieve optimum energy and power densities. The LISC was cycled between 1.5 and 4.5 V, achieving a gravimetric energy density of 103.8 W h/kg and a volumetric power density of 111.8 W h/L. The capacitor retained 85% of its initial capacity after 10,000 cycles. To obtain the good electrochemical performance, a special procedure was performed—the LISC was operated, charged to a given voltage, and then relaxed at open circuit for several cycles. Otherwise, an important fade of the capacity was observed only after

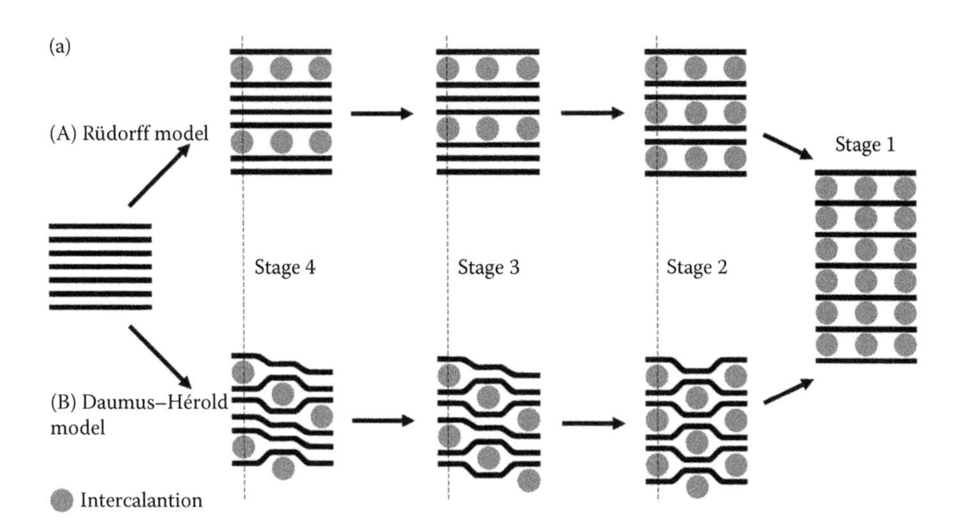

(b) RegionILiC_6 (first stage) ⇆ LiC_{12} (second stage)

RegionIILiC_{12} (second stage) ⇆ LiC_{18} (second stage)

RegionIIILiC_{18} (second stage) ⇆ LiC_{27} (third stage) ⇆ LiC_{36} (fourth stage)

RegionIVLiC_{36} (fourth stage) ⇆ LiC_{72} (eighth stage)

FIGURE 2.2 (a) A schematic of the (A) Rüdorff and (B) Daumas–Hérold models of ion intercalation into graphite [9]; (b) the transition scheme of each stage [9].

500 cycles. The special procedure was to pre-lithiate graphite, and the potential of the graphite anode stabilized around 0.1 V during cycling.

Nowadays, pre-lithiation of carbon anodes is critical for carbon-anode LISCs to achieve high energy density and long cycle life. After pre-lithiation, the potential of carbon anode shifts to ~0 V, which not only widens the potential difference between the negative and the positive electrodes, but also benefits the full usage of the AC cathode. Sivakkumar and Pandolfo constructed a LISC with a pre-doped graphite anode and an AC cathode [11]. The pre-lithiation was terminated once the potential of the graphite electrode dropped from 3 to 0.05 V. They investigated the performance of the LISC at two different cut-off voltages. When the operating voltage was between 3.1 and 4.1 V, the potential of the pre-lithiated graphite rose from 0.13 to 0.2 V after 100 cycles, which can be ascribed to time-dependent loss of stored Li ions in the pre-doped graphite electrode. The potential of the AC positive electrode swung between 3.3 and 4.2 V, corresponding to the adsorption and desorption of anions. However, when the operating voltage was between 2.0 and 4.1 V, the potential of the negative electrode rose from 0.16 to 0.56 V after 100 cycles. The potential of the AC positive electrode swung between 2.2 and 4.2 V, desorbing of anions in the voltage range of 4.2–3.0 V, and adsorbing Li ions in the voltage range of 3.0–2.2 V. For the latter process, the lithium-ions were supplied from the pre-doped graphite electrode. Although the energy density of the LISC increased because the AC electrode swung in a wide potential window, the stability of the LISC was damaged.

For pre-lithiation, metallic lithium is usually assembled into the LISCs to construct the Li/carbon/AC three-electrode cells. Pre-lithiation can be achieved by either externally short circuiting (ESC) the carbon electrode with the metallic lithium or electrochemically (EC) doping the carbon electrode from the metallic lithium. However, metallic lithium might lead to thermal runaway and firing of systems. Decaux et al. constructed a LISC using 2 M lithium bis(trifluoromethane) sulfonimide (LiTFSI) as the electrolyte instead of 1 M $LiPF_6$ [12]. The LISC was in a two-electrode configuration, consisted of a graphite anode and an AC cathode. The pre-doping of the graphite anode was achieved through successive galvanostatic charge/discharge of the two-electrode cell. As the pre-lithiation would consume a large amount of available lithium-ions in the electrolyte solution, LiTFSI salt was used as the electrolyte because of its high solubility. The concentration of LiTFSI could reach 2 M in the organic solvent of EC:DMC (dimethyl carbonate) (1:1, v/v). After a few charge/discharge cycles, the potential of the negative graphite electrode decreased to 105 mV, confirming the formation of second stage graphite intercalation compound (LiC_{12}). If 1 M $LiPF_6$ was used as the electrolyte, the graphite electrode potential could only have reached ~0.5 V (indicating very low lithium intercalation degree) due to low lithium salt concentration in the organic solvent. The constructed LISC could be charged/discharged in the voltage range of 1.5 and 4.2 V. During the galvanostatic cycling of LISC, potential variation of the graphite electrode was very low. The LISC delivered an energy density of 80 W h/kg, which was four times higher than the value for symmetric EDLCs.

In the Li/carbon/AC three-electrode cells, a porous polyolefin film of about 20 mm thick is used to separate the carbon anode and the metallic lithium auxiliary electrode. During the pre-lithiation, metallic lithium is oxidized to release lithium ions into the electrolyte, while reduction of the lithium-ions occurs at the surface of the carbon anode. Under these conditions, the pre-lithiation processes require a very long pre-doping time (>10 h) to reach the sufficient and uniform lithium pre-doping level because of the low diffusion rate of lithium-ions in the electrolyte solution, which is greatly influenced by the distance (20 mm) between the metallic lithium and the carbon electrode [13]. Kim et al. explored an internal short (IS) approach for pre-lithiation by directly attaching a lithium metal to the graphite electrode without the separator and the electrolyte [14]. Due to the direct attachment, lithium-ions had very short travel distance (almost zero). The IS approach method can provide a much faster lithium pre-doping kinetics with the potential of the graphite electrode changing to 4 mV within 1 min. A LISC was constructed using IS pre-lithiated graphite as the anode. The LISC exhibited better electrochemical performance than the LISC based on ESC pre-lithiated graphite and the LISC based on EC pre-lithiated graphite. Ahn et al. constructed a LISC with high oriented graphene sponge (HOG) as the anode and AC as the cathode [15]. HOG was pre-lithiated by the IS approach. The constructed LISC could achieve an energy density of 108 W h/kg at the power density of 1.4 kW/kg with the capacity retention of 84.2% after 1000 cycles.

Park et al. used in-situ synchrotron wide-angle X-ray scattering (WAXS) technique to study the IS pre-lithiation process [16]. On the WAXS curves, there are four discrete (002) peak zones: Q1, Q2, Q3, and Q4, corresponding to graphite→LiC_{24}, LiC_{24}→LiC_{18}, LiC_{12}, and LiC_6, respectively. During the IS process, the intensity of the Q1 peak decreased within 14 min. The Q2 peak appeared in <3 min, reached maximum at 22 min, and disappeared at 49 min. the Q3 and Q4 peaks emerged at 13 min and 28 min, respectively. During ESC and EC processes, however, the Q4 peak was not observed, and the Q2 and Q3 peaks both emerged much later than in the IS process. These were the direct evidences that the IS approach provided much faster pre-lithiation kinetics than ESC and EC methods do.

Park et al. proposed another new pre-lithiation method that used a stable lithium metal oxide; Li_2MoO_3 as the lithium source [17]. Li_2MoO_3 has a typical rhombohedral symmetry with *d* spacing of ~0.25 nm [18], indicating that Li_2MoO_3 can release lithium-ions very well. Li_2MoO_3 was integrated into the AC cathode, and the LISCs were galvanostatically charged up to 4.7 V at a constant current to extract lithium-ions from the Li_2MoO_3 and to insert the lithium-ions into the graphite anode. Due to the irreversibility, <30% of lithium-ions could be recovered to the host structure above 2.5 V (the LISC was charged/discharged between 1.5 and 3.9 V, and the potential of AC cathode swung between 2.5 and 4.0 V), which means that most of the lithium-ions would stayed in the graphite anode. After lithium-ion extraction, the de-lithiated phase of $Li_{2-x}MoO_3$ in the cathode could contribute little to the electrochemical reaction of the LISC, and thus almost had no effect on the performance of the LISC. The pre-lithiation level of the graphite anode could be controlled through controlling the Li_2MoO_3 amount included in the cathode. Compared to the LISC pre-lithiation using metallic lithium, the LISC pre-lithiation with Li_2MoO_3 exhibited much higher capacity, higher Coulombic efficiency, and higher rate capability, which suggests that it has great potential.

Sivakkumar et al. investigated the effect of ball milling on the performance of the graphite battery-type anode LISCs [19]. Ball milling of graphite resulted in a decrease in discharge capacity at relatively low charge rate, while an increase in discharge capacity at relative high charge rate, which was attributed to the interplay of two different charge storage mechanisms: lithium-ion intercalation and lithium-ion adsorption. Pristine graphite material stored the majority of charge mainly via a lithium-ion intercalation. The ball-milled graphite stored charge by lithium-ion adsorption in addition to lithium-ion intercalation because of the increase in its surface area after ball milling. At the low C-rate, the charge stored via ion adsorption was lower than that of ion intercalation. However, at the high C-rate, the charge stored via intercalation dropped rapidly due to the incomplete charging. However, ion adsorption could provide additional charge storage for ball-milled graphite.

Lee et al. used hydrogen peroxide treated graphite as the anode material [20]. After hydrogen peroxide treatment, there were many oxygen-containing groups on the graphite, which reacted with lithium-ions and formed lithium carboxylic salts. The lithium carboxylic salts could bond to the solid electrolyte interface [SEI] layers generated from electrolyte reduction, improving the stability of the SEI layers. The stable SEI layer blocked the co-intercalation of bulky counter ions such as PF_6^-, and thus prevented the exfoliation of exposed edge planes. Therefore, the LISC based on pre-lithiated hydrogen peroxide treated graphite anode exhibited improved rate capability and stable capacity retention.

2.1.2 Hard Carbon

Another type of carbon material that could be used for the battery-type anodes in LISCs is HC, which is formed with interlaced single-layer graphite nanosheets. Since lithium-ions can be adsorbed on both sides of the single-layer graphite sheet, HC has a higher specific lithium storage capacity than graphite. Moreover, the large space gap between the carbon layers of HC is beneficial for the lithium-ion intercalation/deintercalation, which is particularly desirable for power sources. Kim et al. compared the electrochemical performance of graphite and HC and found that HC showed the better rate capability and longer cycle life [21]. Ni et al. studied the charge/discharge behavior of HC using a HC/Li half-cell [22]. They found that lithium-ions could not only insert/extract between the carbon layers of HC but also adsorb/desorb in both sides of the single sheets or walls of nanopores [23]. The HC delivered a capacity in a wide potential range and achieved a capacity of 526 mAh/g with the efficiency of 80%. A LISC prototype was constructed in 1 M $LiPF_6$/EC:DMC (1:3, v/v) using HC and AC as the anode and cathode, respectively. The LISC was cycled between 2.8 and 4.0 V, achieving an energy density of 20.8 W h/kg at the power density of 480 W/kg, and maintained 90% of its initial capacity after 3000 cycles. The LISC also displayed good rate capability with the capacity changing from 21.5 to 20.7 F/g when the discharge current increases from 100 to 400 mA/g.

Zhang et al. introduced two kinds of HC, i.e., spherical HC and irregular HC, to assemble LISCs [24]. Before the assemblies, either spherical HC or irregular HC electrode was pre-lithiated with metallic lithium, and the potentials of the HC anodes were stabilized close to 0 V with very small variations during the charge/discharge of the LISCs. Similar to the pre-lithiation of graphite anode, the pre-lithiation of

HC anode can greatly improve the electrochemical performance of the LISCs. XRD results showed that the interlayer spacing of irregular HC was larger than that of spherical HC, which is beneficial for intercalation/deintercalation of lithium-ions. Thus, the LISC using irregular HC as the anode material exhibited a better rate performance. Moreover, an obvious potential plateau below 0.1 V was observed on the charge/discharge curves of the irregular HC, which was not observed on the charge/discharge curves of spherical HC. The potential plateau corresponded to the intercalation/deintercalation of lithium-ions into the micropores among the stacked graphite layers. When the LISCs were cycled between 2.0 and 4.0 V, the potential of the spherical HC electrode varied between 0.042 and 0.197 V. However, due to the existence of the potential plateau below 0.1 V, the irregular HC electrode varied only between 0.021 and 0.14 V, leading to the improved utilization of the AC positive electrode. Therefore, the LISC assembled with the irregular HC anode displayed better electrochemical performance, delivering an energy density of 85.7 W h/kg and a power density of 7.6 kW/kg and retaining about 96.0% of its initial capacity after 5000 cycles at 2 C rate.

Cao et al. developed a LISC based on a HC anode covered with a layer stabilized lithium metal powder (SLMP) film (Figure 2.3) [25]. The SLMP film decreased the anode potential from 3.0 to 0.47 V, and the OCV of the LISC reached 2.9 V. When the LISC was cycled between 2.0 and 3.9 V, the anode swung between 0.12 and 0.47 V, and the AC cathode electrode swung between 2.27 and 4.02. The LISC achieved an energy density of about 25 W h/kg based on the weight of the cell with good cycling stability and rate capability.

2.1.3 Graphene

Graphene is another unique carbon allotrope that is the single layer of carbon atoms in a honeycomb lattice structure. Graphene has outstanding mechanical properties, thermal and electrical conductivities, high carrier mobility, and large specific surface area. It is reported that graphene nanosheets have theoretical capacity of 744 mAh/g, which corresponds to insertion of one lithium-ion per three carbon atoms [23,26]. Both Ren et al. [27] and Acznik et al. [28] reported their work on assembly of LISCs using pre-lithiated reduced graphene oxide and AC as the anode and cathode, respectively. However, the high irreversible capacity and large voltage hysteresis limited the application of graphene in LISCs. Sun et al. fabricated a single-wall carbon nanotube/graphene (SG) composite and constructed a LISC using the composite as both anode and cathode materials [29]. The pristine composite was used as the EDLC cathode, while the pre-lithiated SG composite was used as the battery anode. Due to the SWCNT spacers, the SG composite showed good rate capability, displaying a specific capacity of 145 and 101 F/g at 0.05 A/g and capacity 1 A/g, respectively. The charge capacity of the pre-lithiated SG electrode reached 503 mAh/g because lithium-ions could be adsorbed on both sides of the nanosheets. The SWCNTs not only enlarged the interspace of graphene nanosheets but also prevented graphene from re-stacking during cycling. The LISC exhibited an energy density of 222 W h/kg at a power density of 410 W/kg. After 5000 cycles, the Coulombic efficiency was maintained at 97%, and the capacity was maintained at 58%.

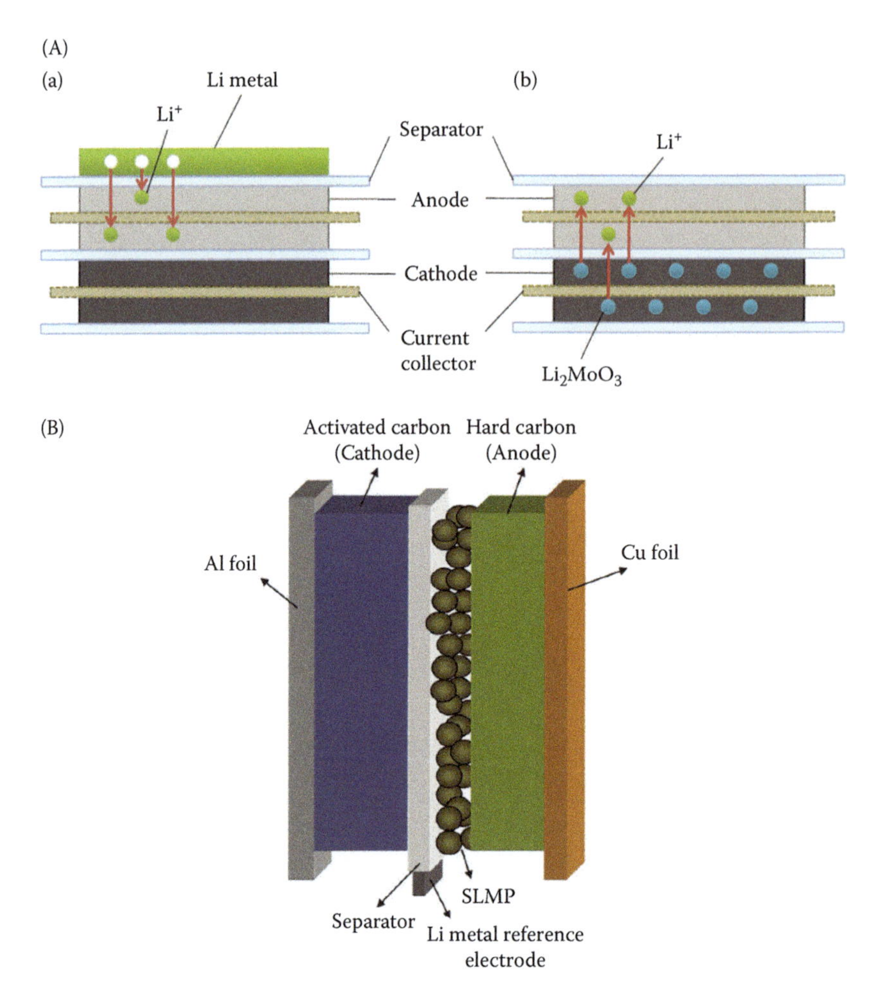

FIGURE 2.3 (A) Schematic diagrams showing different lithium-doping methods: (a) conventional lithium doping using metallic lithium in the cell and (b) proposed method using Li_2MoO_2 as an alternative lithium source [17]; (B) a schematic diagram of an AC/SLMP surface applied HC LISC configuration [25].

2.1.4 Mesocarbon Microbeads

Zhang et al. constructed a LISC using pre-lithiated mesocarbon microbeads (MCMB) as the anode material to investigate the effect of the pre-lithiation degree on the electrochemical behavior of the LISC [30]. When the pre-lithiation capacity was lower than 200 mAh/g, the MCMB maintained its graphitic structure. However, when the pre-lithiation capacity increased to 250 mAh/g, LiC_{24} and LiC_{12} were formed. When the pre-lithiation capacity continued to increase, LiC_6 could be observed. The constructed LISC was cycled between 2.0 and 4.0 V, and it was found that the charge/discharge profile of pre-lithiated MCMB showed a typical U shape of a graphite electrode when the pre-lithiation capacity was lower than 200 mAh/g.

When the pre-lithiation capacity reached 250 mAh/g, the potential of the MCMB anode varied only between 0.05 and 0.08 V (ΔV=0.11 V). The reduced potential variation range fully benefited from using the capacity of the AC cathode, and thus improved the electrochemical performance. The LISC with the pre-lithiation capacity of 300 mAh/g showed the maximum energy density of 92.3 W h/kg and the highest power density of 5.5 kW/kg. They also found that introducing HC into the MCMB anode can greatly improve the electrode's electronic conductivity, and thus improve the rate performance of the LISC [3].

2.1.5 Soft Carbon

A soft carbon, petroleum coke (PeC), was used by Schroeder et al. Self-discharge of the pre-doped graphite is another important factor that affected the performance of the LISC. The Self-discharge arises from the direct contact between graphite and electrolyte, which could be avoided by uniform SEI layer. Sivakkumar and Pandolfo tried three methods for pre-lithiation. They found that pre-lithiation method that formed a uniform SEI layer could effectively prevent lithium-ions from leaking, and the assembled LISC could be operated between 2.0 and 4.1 V, delivering an energy density of 100 W h/kg [31].

Besides carbons, metal oxides are also very promising for their usage as battery-type anodes in LISCs because of their non-toxicity, good chemical stability, low solubility, and excellent safety.

2.2 TITANIUM OXIDES

2.2.1 TiO_2

The insertion/de-insertion of lithium-ions into TiO_2 takes happen at ~1.8 V, which lies within the stable potential window of liquid electrolytes [32]. Based on the fully intercalated composition of $LiTiO_2$, the theoretical capacity of TiO_2 is 336 mAh/g [33]. However, the low ionic conductivity (~1.2×10^{20} /Sm^2) and the poor electronic conductivity (~1×10^{-12} S/m) have limited the LISCs consisted of TiO_2 battery-type anode to achieve high power [34]. To improve the electrochemical performance of TiO_2 anode, mesoporous and nano-sized TiO_2 were fabricated for the reduction of lithium-ion diffusion distance and the accommodation of strain during cycling, and TiO_2/carbon composites were produced to improve the conductivity.

Cho et al. synthesized mesoporous anatase TiO_2 spherical particles using dodecylamine to generate pores within particles as a structure directing agent to improve the mobility and chemical diffusion of lithium-ions in the anatase TiO_2 [35]. The diameter of the obtained spherical particles was ~400 nm with the pores in the spherical particles around 10 nm. The half-cell test showed that the mesoporous anatase TiO_2 spherical particles exhibited good rate ability, delivering charge capacities of about 220 mAh/g at 0.1 C and 160 mAh/g at 5 C. A LISC was built using the spherical particles and AC as the anode material and cathode materials, respectively. A specific capacity of 23 F/cm^3 could be achieved at the current density of 0.5 mA/cm^2. However, due to the poor electronic conductivity of TiO_2, the LISC only maintained 91% of its initial capacity after 100 cycles.

Cai et al. also reported the synthesis of mesoporous TiO_2 microspheres through a facile template-free solvothermal method followed by a heat treatment [36]. The calcination temperature had great effect on the electrochemical performance of the mesoporous TiO_2 microspheres. XRD patterns showed that the samples calcined at low, medium, and high temperatures composed of anatase and TiO_2(B), pure anatase, and anatase and rutile, respectively. Such a phase transition may have an effect on the electrode performance. The specific capacity of the TiO_2 microspheres decreased with the increasing calcination temperature because the specific surface area of the mesoporous TiO_2 microspheres decreased with the increasing calcination temperature and rutile TiO_2 is relatively inert to lithium intercalation. A LISC was assembled using the mesoporous TiO_2 microspheres calcined at 400°C as the anode material and AC as the cathode. The LISC achieved a very high power density of 9.45 kW/kg while maintaining an acceptable energy density of 31.5 W h/kg. The LISC retained 98% of its initial capacity after 1000 cycles.

Wang et al. constructed a LISC whose anode was made of free-standing TiO_2 nanobelt arrays (TNBAs) grown on Ti foil without any ancillary materials [37]. Since TNBA was directly used as the anode without any conductive carbon and polymeric binder, it possessed efficient transport pathways for both lithium-ions and electrons. TNBA showed good rate capability (delivering a capacity of about 160 mAh/g at the current density of 0.1 A/g and a capacity of 43 mAh/g even at 5 A/g), cycling stability (maintaining 86% of its initial capacity after 100 cycles at the current density of 0.2 A/g), and reversibility (charge/discharge curves did not significantly change during cycling). The cathode of the LISC was made of graphene hydrogel of 3D porous network, and the specific capacity value for graphene hydrogels was as high as 133 F/g at 0.5 A/g. The LISC based on the TNBA anode and the graphene hydrogel cathode could deliver the maximum energy density of 82 W h/kg, and achieved an energy density of 21 W h/kg at a high power density of 19 kW/kg.

It was found that surface area is another important factor for the electrochemical performance of TiO_2 because the large surface area is favorable for smooth insertion/extraction reactions between the electrolyte and anode. Kim et al. and Choi et al. reported the constructions of high-performance LISCs using urchin-like TiO_2 anode and AC cathode [38,39]. Urchin-like TiO_2 was composed of numerous compact nanorods and therefore has a larger BET-specific surface. The specific area of urchin-like TiO_2 first increased with the reaction time until 20 h, and then decreased with the reaction time because of the broken nanorods derived from excessive growth. The specific capacity of urchin-like TiO_2 was closely related to the specific surface area, increasing to a maximum at 20 h and then decreasing. Therefore, the LISC exhibited a power density of 194.4 kW/kg and an energy density of 50.6 W h/kg and maintained 89% of its initial capacity after 5000 cycles [38]. Choi et al. compared the electrochemical performance of the LISC based on urchin-like TiO_2 anode with the LISCs based on other TiO_2 anodes (including anatase TiO_2 nanoparticle, TiO_2 nanorod, flower-like TiO_2), and found that the urchin-like TiO_2 has the smallest R_{ct} value and the largest lithium-ion diffusion coefficient because the large surface area and the porous structure of the urchin-like TiO_2 increased the electrolyte–electrode contact area for the lithium-ion insertion/extraction and reduced the transport path for Li^+ and electrons [39].

Bauer et al. prepared Mo⁻ and Nb-doped anatase nanomaterials, $Mo_{0.1}Ti_{0.9}O_2$ and $Nb_{0.25}Ti_{0.75}O_2$, which are used as the anode materials in LISCs [40]. Nb^{5+} and Mo^{6+} ions were doped into the anatase structure and replaced Ti^{4+} in the TiO_6 octahedra, producing higher valence state, which improved the pseudocapacitive charge storage in these materials. The LISC that is composed of the doped TiO_2 anode and AC cathode could achieve an energy density of 41 W h/kg at 1.7 kW/kg and an energy density of 36 W h/kg at 3.2 kW/kg.

Chen et al. mixed hydrophobic TiO_2 nanocrystals (NCs) with hydrophilic carbon nanotubes (CNTs) in nonpolar solvent [41]. Upon addition of polar solvent, solvation forces induce the hydrophobic TiO_2 NCs to assemble around the hydrophobic CNTs, forming CNT-threaded microspheres. After calciniation, the microspheres converted to robust CNT-threaded porous particles. The diameter of the particles ranged from a few hundred of nanometers to a few micrometers. In the particle, the highly crystalline anatase TiO_2 NCs were of ~5 nm. Meanwhile, the CNTs (with diameter of 20–30 nm and length of 5–10 μm) were intertwined, forming a hierarchically porous structure with both macropores and mesopores, which benefited electrolyte transportation. On the charge/discharge plots of the composite electrode, a distinct voltage plateau was observed, and a capacity of ~130 mAh/g was estimated from the plateau region. The composite electrode could deliver a capacity of 270 mAh/g because another capacity of 170 mAh/g was estimated from the slop region of the charge/discharge plot. In addition, the electrode exhibited good rate capability, delivering a reversible capacity of about 265 and 220 mAh/g at rates of 1 and 20 C, respectively. The superior electrochemical performance could be ascribed to the following: (1) The CNTs in the composite increased the conductivity; (2) the hierarchical porous structure formed by the CNTs benefited electrolyte transportation; and (3) small size and high surface area of TiO_2 provided abundant surface active sites for fast surface/interfacial lithium reactions. A LISC was assembled using the composite anode and AC cathode, which could provide an energy density of 59.6 W h/kg at a power density of 120 W/kg.

Gao et al. reported on the construction of rutile TiO_2-decorated hierarchical LTO (RLTO) nanosheet arrays on Ti foil [42]. The obtained self-supported arrays avoided using a polymeric binder, and thus possessed faster lithium-ion and electron transportation. Transmission electron microscopy (TEM) images showed that the RLTO nanosheets were composed of nanorods and nanoparticles, which formed the hierarchical and interconnected architecture containing both macropores and mesopores. Macropores around 1 mm served as the reservoirs for electrolytes, and mesopores around 5–10 nm provided accessible channels for lithium-ions. Lattice fringes belonging to [011] direction of LTO and [001] direction of rutile TiO_2 could be found on the high-resolution transmission electron microscopy (HRTEM) images, and both of the directions are beneficial to lithium-ion transportation. The RLTO nanosheet array electrode displayed good rate ability (a discharge capacity of 185 mAh/g at 1 C and a discharge capacity of 142.9 mAh/g at 30 C), cycling stability (maintaining 92.3% of its initial capacity after 3000 cycles at 30 C), and high Coulombic efficiency (close to 100%). Moreover, even after 3000 cycles at 30 C, no obvious structural change could be observed. A LISC was assembled by utilization of the self-supported RLTO arrays as the anode and nitrogen-doped nanotubes

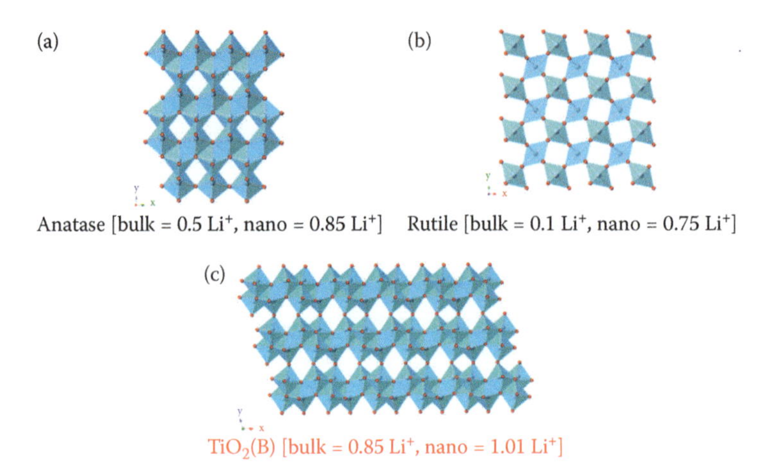

FIGURE 2.4 Unit cells of anatase (a), rutile (b), and TiO_2(B) (c) with idealized lithium-ion insertion sites [43].

coated Al foil as the cathode. The LISC exhibited an ultrahigh energy density of 74.85 W h/kg at a power density of 300 W/kg (Figure 2.4).

TiO_2(B) has a monoclinic C2/m structure comprised of edge and corner sharing TiO_6 octahedra with an open channel parallel to the b-axis that sits between axial oxygen molecules [43]. Though the unit cell of TiO_2(B) contains 8 Ti sites and can theoretically accommodate 10 Li ions, only 8 Li ions can be filled because of the repulsion between Li ions [44]. Calculation results reveal that in TiO_2(B), lithium-ions can diffuse through low-energy pathways instead of bulk diffusion-controlled mechanism [43]. Thus, TiO_2(B) shows good performance even at high current densities, leading to the high rate ability.

Aravindan et al. explored monoclinic TiO_2(B) as the insertion type anode for LISC [45]. Monoclinic TiO_2(B) nanorods were synthesized using commercial anatase TiO_2 powder via conventional hydrothermal method and a subsequent ion-exchange reaction with protons. The obtained monoclinic TiO_2(B) was of a tunnel-like structure, which was favorable for lithium-ion diffusion. Moreover, the TiO_2(B) nanorods form a highly interconnected network, which may trap a large amount of liquid electrolytes and provide facile lithium-ion transportation. Half-cell test showed that the TiO_2(B) nanorods exhibited higher reversible capacity (>0.5 mole of lithium-ions per formula unit) and lower insertion potential (~1.5 V) than its anatase TiO_2 phase. In addition, TiO_2(B) nanorods showed good cycling stability, retaining 93% of its reversible capacity after 200 cycles. A flexible LISC was assembled using TiO_2(B) nanorods and AC as the anode and cathode materials, respectively. The cell exhibited a maximum energy density and power density of 23 W h/kg and 2.8 kW/kg, respectively, and the LISC retained 73% of its initial capacity after 1200 cycles at a current density of 1.5 A/g.

Wang et al. assembled a LISC using TiO_2(B) nanowires and CNT as the anode and cathode materials, respectively [46]. The LISC was operated over a wide voltage range between 0 and 2.8 V, delivering an energy density of 12.5 W h/kg at even a rate of 10C. Despite the capacity loss in the first hundred cycles, the LISC maintained desirable cycling stability. The irreversible capacity loss might be ascribed to the reactive Ti–OH and Ti–O surface sites that can cause unwanted electrolyte degradation and irreversible trapping of lithium-ions.

Though lithium-ions can diffuse in TiO_2(B) through low-energy pathways instead of bulk diffusion-controlled mechanism, lithium-ion diffusion only proceeds along the *b*-axis tunnel, giving a poor lithium-ion diffusion coefficient. Usually, TiO_2(B) synthesized via hydrothermal treatment from alkaline titanates gives cylindrical morphology with long *b*-axis, resulting in poor C-rate capability. Naoi et al. prepared nanosized-TiO_2(B)/MWCNT composite through 0 combined with a follow-up hydrothermal treatment [47]. The TiO_2(B) nanoparticles were of 5 nm in average size, and anisotropic crystal growth with ultrashort *b*-axis. Due to the ultra-centrifuged treatment, TiO_2(B) nanoparticles were homogeneously dispersed into the MWCNT matrix. The TiO_2(B)/MWCNT delivered a capacity of 275 mA/g at 1 C and 235 mAh/g at 300 C. Such a superior electrochemical performance was owing to the shortening of *b*-axis length and increased number of diffusion paths, which enabled a fast lithium-ion access and intercalation into TiO_2(B) A1 and A2 sites along the *b*-axis diffusion channel. The MWCN not only increased the electronic conductivity the composite but the mesopores within the MWCNT matrix also acted as a reservoir of lithium-ions for high-rate charge/discharge operation. It is believed that the uc-TiO_2(B)/MWCNT composite can be used to construct LISCs of high electrochemical performance. Liu et al. prepared mesoporous TiO_2(B) microspheres [48]. The microspheres showed high specific capacity, high rate ability, and superior cycling stability, which is also favorable for the construction of high-performance LISCs.

2.2.2 $Li_4Ti_5O_{12}$

LTO is a kind of lithium titanium oxide of spinel structure. In LTO, all the tetrahedral 8a sites are occupied by lithium and the octahedral 16d sites are shared by lithium and titanium with an atomic ratio of 1:5 in the cubic oxygen array [49]. After lithiation, three lithium atoms at the 8a sites move to the empty 16c sites and the new lithium atoms also take up the 16c sites, converting to a rock salt structure $Li_7Ti_5O_{12}$ [50]. The variation of lithium content and occupying sites during lithiation/delithiation was also observed by Lu et al. using the spherical aberration-corrected scanning transmission electron microscopy in combination with theoretical calculations [51]. In addition to variation of lithium content and occupying sites, it was found that titanium ions also adopted locally different valences upon lithiation.

Although the theoretical capacity of LTO is only 175 mAh/g, it still receives much attention in LISC because the coulombic efficiency of LTO reaches >95% at 1 C as well as its low cost and easy synthesis [52]. The other advantages of LTO being a lithium insertion host include the little volume change during charge/discharge because of the zero-strain insertion, the flat charge/discharge plateau at ~1.5 V avoiding the electrolyte decomposition and SEI formation, and the smaller heat generated during charge/discharge because of the smaller entropy change [53]. However, similar to TiO_2, the application of LTO in LISC is also limited by its low lithium-ion diffusion coefficient and its poor electronic conductivity. Therefore, nano-sized and porous LTO was thus synthesized to short the pathways for solid-state diffusion of Li ions [54,55].

Lee et al. prepared granule LTO by spray drying process with precursor slurry of Li_2CO_3 and TiO_2 and followed by calcination in air at 800°C for 6h. The granule LTO was of porous structure, composed of primary particles with the size ranging from 300 to 500nm. A LISC was assembled using the granule LTO and AC as the anode and cathode, respectively [56]. Since unexpected gassing is one of the main factors to decrease cycling performance, the gassing of the LISC was investigated at different current densities. H_2 and HF were the main components of the unexpected gas, which resulted from decomposition of the electrolyte, and F element could be detected in the LTO electrode due to the decomposition. Thus, the electrolyte decomposition extent could be evaluated through the measurement of the F element concentration in the LTO electrode. It was found that the F element concentration in the LTO electrode at current density of 1A/g showed about two times higher than that of 0.8A/g because of the acceleration of gas evolution at high current density. The accelerated gassing had a negative effect on the cycling stability of the cell. The capacity retention reached 97.3% at a current density of 0.8A/g after 1000 cycles, while the capacity retention decreased to 83.7% at a current density of 1A/g.

Although nano-sizing and porosity are effective ways to overcome the low lithium-ion diffusion coefficient of LTO, the electrochemical performance of LTO is still restricted by its poor electric conductivity. Naoi et al. prepared a novel composite, nano-crystalline LTO grafted on carbon nano-fibers (nc-LTO/CNF) [53]. In the composite, the LTO particles had a diameter of ~5–20nm and were well dispersed among the CNFs. On the charge/discharge curves, the dominant plateau of nc-LTO/CNF was observed at ~1.5V, which is the same as LTO. As the nc-LTO facilitated lithium-ion diffusion, and the LTO/CNF junction improved electronic conductivity, the composite displayed good rate capability and cycling stability. The nc-LTO/CNF composite showed a reversible capacity of 158mAh/g per LTO even at 300C and maintained 90% of its initial capacity after 9000 cycles at 20C. The LISC consisting of the nc-LTO/CNF anode and AC cathode showed a high energy density of 40W h/L and a high power density of 7.5kW/L.

Choi et al. synthesized LTO-activated carbon hybrid nanotube (LTO/ACNT) via electrospinning followed by carbonization and thermal activation [57]. The carbon in the nanotube improved the conductivity of the hybrid, leading to the decrease of the charge transfer resistance (R_{ct}) and electrolyte resistance (R_s). Furthermore, the ACNT also shortened the lithium-ion diffusion distance. The LTO/ACNT composite exhibited good rate capability, achieving a specific capacity of 128mA/g at a rate of 100mA/g, and a specific capacity of 84mA/g at the rate of 4000mA/g. A LISC was constructed using LTO/ACNT as the negative electrode material and AC as the positive electrode material with the mass ratio of 3:1. The cell was cycled between 0.5 and 3.5V and exhibited a high energy density in the range of 90–32W h/kg over the power densities from 50 to 6000W/kg. Choi et al. also reported in one of their work on preparation of hyper-networked LTO-N-enriched carbon hybrid nanofiber sheets (LTO-CHNS) using the electrospinning method [58]. LTO-CHNS also exhibited superior rate capability, delivering a specific capacity of 135mAh/g even at the charge/discharge rate of 4000 mA/g. A LISC composed of LTO-CHNS anode and AC cathode exhibited a high energy density of 91 and 22 W/kg at power densities of 50 and 4000 W/kg, respectively. The excellent electrochemical performance of the LISC could be ascribed to the presence of an amorphous carbon phase in the hybrid nanosheet, which enhanced the electron transport kinetics because of the continuous electron pathway as well as providing additional capacity at high charge/discharge rates simultaneously.

Kim et al. obtained graphene-wrapped LTO particles via a one-step process by adding LTO particles into the mixture of NH_4OH, NH_2NH_2, and graphene oxide solution [4]. TEM images showed that the surface of the LTO particles was covered by the graphene layer of 3 nm thick. The wrapping LTO particles with graphene helped to remedy the intrinsically low electronic conductivity of LTO, and dramatically improved the electrochemical performance of LTO. The graphene-wrapped LTO particles exhibited a specific capacity of 150 mAh/g with a Coulombic efficiency close to 100% up to 20 cycles. A LISC was constructed by using the graphene-wrapped LTO anode and an AC cathode. The specific capacities were 83.6 and 25.7 F/g at 0.1 and 15 mA/cm^2, respectively, indicating the good rate ability. The LISC maintained 75% of its initial capacity after 1000 cycles, indicating the good cycling stability. The LISC could deliver a high specific energy of up to 50 W h/kg and could even maintain an energy of ~15 W h/kg at a charge/discharge rate of 20 s.

Self-supported electrode, which avoids the conventional slurry-processed electrodes using a polymeric binder, can possess faster lithium-ion and electron transportation. Zuo et al. prepared a thin-film LISC with both high volumetric energy and power densities. Different with the slurry-processed electrodes, both the cathode and anode were binder-free with the active nanomaterials growing directly on carbon cloth current collector [59]. The LTO electrode was prepared by converting the rutile TiO_2 nanowire array on carbon cloth into the LTO nanowire array. The LTO array electrode could deliver capacity of 0.235 mAh/cm^2 and retain 95% of the initial value after 400 cycles. Thin-film LISC was assembled using the LTO nanowire array as anode and MWCNT array as cathode. The LISC exhibited high volumetric energy comparable to commercial thin-film lithium batteries and high power densities comparable to commercial thin-film lithium supercapacitors. Moreover, the LISC device showed good cycling stability of more than 3000 times, which is extremely attractive for thin-film downsized energy storage system (Figure 2.5).

2.2.3 $LiCrTiO_4$ Spinel

$LiCrTiO_4$ (*s*-$LiCrTiO_4$), which can be obtained by replacing Ti atoms in LTO with Cr atoms on the corresponding 16d sites, is another promising battery-type anode material. Since the doping of Cr into Ti lattice leads to the overlap of Cr 3*d* bands with Ti 3*d* bands, *s*-$LiCrTiO_4$ exhibits high electronic conductivity [61,62]. The theoretical capacity of *s*-$LiCrTiO_4$ is 157 mAh/g, close to that of LTO. On the cyclic voltammograms of Li/$LiCrTiO_4$ half-cells (cycled between 1.0 and 3.0 V vs. Li), a sharp pair of redox peaks belonging to $Ti^{4+/3+}$ redox couple could be observed. The small potential difference of the redox peaks implied reversible insertion and extraction of lithium

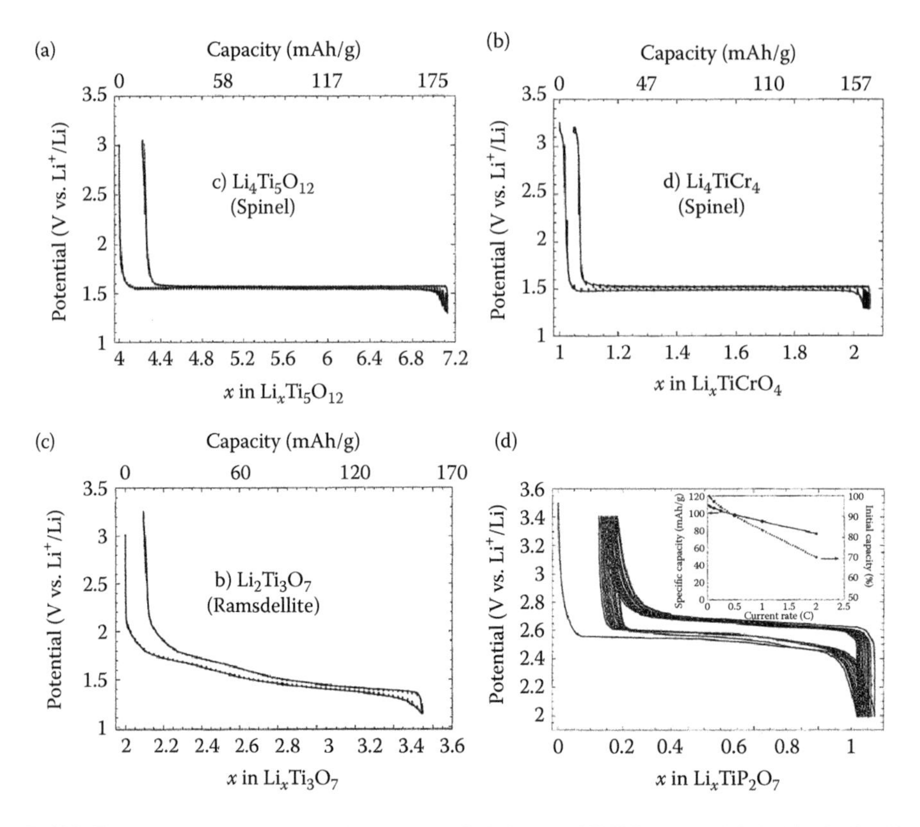

FIGURE 2.5 Potential-composition curves of LTO: (a), $LiCrTiO_4$ (b), $Li_2Ti_3O_7$ (c) obtained in galvanostatic intermittent mode: regime of C/20 for 30 min and relax period of 30 min, and (d) Charge/discharge profile under a galvanostatic mode at current density of C/10 for TiP_2O_7 [60].

ions in *s*-$LiCrTiO_4$ lattice [61]. The charge and discharge curves of Li/$LiCrTiO_4$ half-cell cells showed only one plateau, and the potentials of the plateaus were all above 1.5 V, and thus Li plating and SEI layer formation were avoided [61,63]. Rao and Rambabu constructed a LISC using nanocrystalline *s*-$LiCrTiO_4$ as anode and AC as cathode [64]. The LISC exhibited a specific capacity of 59 mAh/g, and 96% capacity retention up to 1000 cycles. The nanosize of the crystal made shorter distance for lithium-ion diffusion, which also contributed to the electrochemical performance of the LISC. Aravindan et al. exploited a LISC using sub-micrometer-sized *s*-$LiCrTiO_4$ as anode and AC as counter electrode. By optimizing the active mass ratio, the specific capacity of the LISC improved to 83 F/g and maintained 85% of its initial capacity after 1000 cycles [65].

2.2.4 Other Titanium Compounds

Other titanium compounds can be used as the insertion battery type anode in LISCs include $LiTi_2(PO_4)_3$ and TiP_2O_7. In $LiTi_2(PO_4)_3$ and TiP_2O_7, however, the insertion/de-insertion of lithium-ions takes place at ~2.5 V, which greatly narrows the operation voltage range of LISCs. Therefore, there are only a few reports on the construction of LISCs using $LiTi_2(PO_4)_3$ or TiP_2O_7 as the anode materials.

$LiTi_2(PO_4)_3$, adopting the so-called NASICON-type (sodium superionic conductor) phase, has shown high reversibility for lithium-ion intercalation and deintercalation. The crystal structure of $LiTi_2(PO_4)_3$ exhibits a three-dimensional framework, which consists of corner-sharing PO_4 tetrahedra and TiO_6 octahedra. Each PO_4 tetrahedron is connected with four TiO_6 octahedral units and, conversely, each TiO_6 octahedra is connected with six PO_4 tetrahedra units [66]. There are two types of cavities, M1 and M2, in the Nasicon structure of $LiTi_2(PO_4)_3$. The M1 cavity is situated between two TiO_6 octahedral units and is occupied by a Li-ion. The M2 cavity comprises distorted eightfold coordination and is surrounded by M1 interstitial voids. In $Li_3Ti_2(PO_4)_3$, the three Li ions are reported to occupy two M3 sites and one M′3 site within the M2 cavity, respectively. It is believed that $LiTi_2(PO_4)_3$ can reversibly insert two lithium-ions according to a two-phase mechanism between $LiTi_2(PO4)_3$ and $Li_3Ti_2(PO4)_3$ [60]. A pair of sharp redox peaks belonging to the $Ti^{4+/3+}$ redox couple could be observed on the cyclic voltammogram of $LiTi_2(PO_4)_3$ cycled between 2.0 and 3.0 V. The potential gap between the redox peaks was narrow, and the charge/discharge profile of $Li/LiCrTiO_4$ half-cell cell showed large and distinct voltage plateau at ~2.4 V [66].

Although the theoretical specific capacity of $LiTi_2(PO_4)_3$ reaches 138 mAh/g, the low intrinsic electronic conductivity has limited its application in LISC hybrid electrochemical capacitor. Carbon coating, which can improve the electronic conductivity of $LiTi_2(PO_4)_3$, is a good way to make $LiTi_2(PO_4)_3$ work efficiently at high current densities. Luo et al. reported their work on the construction of LISCs in Li_2SO_4 aqueous electrolyte, using carbon-coated $LiTi_2(PO_4)_3$ as anode and AC as cathode [67]. The LISC delivered a capacity of 30 mAh/g and an energy density of 27 W h/kg based on the total weight of the active electrode materials. The cell maintained over 85% of its initial energy density after 1000 cycles. Aravindan et al. constructed a $LiTi_2(PO_4)_3$/AC LISC with the carbon-coated nano-sized $LiTi_2(PO_4)_3$ as negative electrode and AC as positive electrode in nonaqueous electrolyte [68]. The LISC could be cycled between 0 and 3 V Li due to the nonaqueous solution. The LISC delivers a maximum energy density of 14 W h/kg and a power density of 180 W/kg.

TiP_2O_7, which showed reversible insertion and extraction of ~0.8 mole of Li per formula unit at the voltage of 2.6 V, has been studied as a Li insertion host. As a poly-anion compound, TiP_2O_7 can keep their crystal structures during lithium-ion intercalation/deintercalation process, leading to excellent thermal stability and cyclic performance [69]. The structure of TiP_2O_7 is composed of diphosphate groups linked to TiO_6 octahedra in a NaCl-type arrangement [60]. Two pairs of sharp redox peaks were observed on the increment capacity voltammograms of TiP_2O_7. The pair of redox peaks at 2.63 V corresponds to a two-phase reaction between TiP_2O_7 (A) and $Li_xTiP_2O_7$ (x~0.5, B), while the redox peaks at 2.57 V probably include a second two-phase domain (B′+C) surrounded by two short solid-solution regions of both extreme compositions (B to B′ and C to C′). XRD results confirmed the presence of two-phase domains and solid-solution regions, highlighting the existence of three kinds of phase (A, B, and C) during the redox processes [69]. Aravindan constructed a LISC using TiP_2O_7 and AC as the anode and cathode materials. The LISC was cycled between 0 and 3 V, and it showed a deliverable maximum of 13 W h/kg and 371 W/kg energy and power density, respectively.

2.3 VANADIUM COMPOUNDS

$Li_3V_2(PO_4)_3$ (LVP) exists in two phases, a monoclinic phase and a rhombohedral phase. The monoclinic structure consists of a slightly disordered VO_6 octahedra and PO_4 tetrahedra linked together via oxygen vertices. Li(1) occupies the tetrahedral site while Li(2) and Li(3) occupy pseudo-tetrahedral sites, which allows fast lithium-ion diffusion kinetics [70]. LVP can behave as both an anode (V^{3+}/V^{2+}) and a cathode ($V^{3+}/V^{4+}/V^{5+}$). Satish et al. fabricated LVP coated with amorphous carbon (LVP-C) to increase the electronic conductivity of LVP [71]. The LVP-C showed three peaks on the cyclic voltammogram between 1 and 3 V. The peaks at ~1.9 and ~1.83 V corresponding to the insertion of the first Li. The peak was separated because of the existence of a stable $Li_{3.5}V_2(PO_4)_3$ phase. The peaks at ~1.6 V indicated the insertion of the second lithium-ion. LVP-C has a capacity of 92 mAh/g. It retains 80% of its initial capacity after 100 cycles. Meanwhile, LVP-C exhibits an excellent rate ability, attributed to the good electronic conductivity and stable intercalation structure. A LISC was assembled using LVP-C anode and AC cathode. The LISC displayed an energy density of 22 W h/kg at a power density of 325 W/kg, and retained 66% of its initial capacity after 1000 cycles. The LVP-C could also be used as the cathode electrode material. The cyclic voltammogram of LVP-C between 3 and 4.5 V showed three pairs of redox peaks. Although only two Li ions were extracted from LVP, three pairs peaks are observed due to the existence of a stable $Li_{2.5}V_2(PO_4)_3$ phase. The LISC using AC anode and LVP-C cathode displayed a better performance because the lithium-ion diffusion coefficient of LVP-C operated as a cathode is an order higher than that when it operated as an anode.

Chen et al. developed a nanocomposite film composed of interpenetrating networks of CNTs and V_2O_5 nanowires. The composite film possessed a hierarchically porous structure with large pores enabling rapid electrolyte transport and the small pores increasing the surface area available for electrochemical reactions. In the composite, the small V_2O_5 nanowire dimension allowed effective lithium-ion diffusion. The electronic conductivity of the composite film was improved to 3.0 S/cm by CNTs. The composite film possessed a hierarchically porous structure with large pores enabling rapid electrolyte transport and the small pores increasing the surface area available for electrochemical reactions. The composite showed that an improved electrode kinetics with the intercalation/deintercalation of lithium-ions took place between ~2.8 and 3.0 V. The LISC consisted of the composite anode and AC cathode that could deliver an energy density of 40 W h/kg at a power density of 210 W/kg.

2.4 IRON OXIDES

Metal oxides such as SnO_2, Co_3O_4, and Fe_3O_4 were also considered good candidates as negative electrodes in LISCs due to their large theoretical capacity and relatively low theoretical potential plateau (~0.8 V). Fe_3O_4 is attracting great attention for its high theoretical capacity (924 mAh/g), low cost eco-friendliness, and natural abundances. Zhang et al. designed a LISC using a Fe_3O_4/graphene composite as the anode and graphene-based three-dimensional porous carbon material as the cathode [72]. The Fe_3O_4/graphene composite was prepared by a simple in situ solvothermal method using graphene oxides and $FeCl_3$ as the precursor. In the composite, a large quantity of Fe_3O_4 nanoparticles were anchored onto the graphene nanosheets. The graphene nanosheets not only acted as a good conductive network for ions and electrons but also prevented Fe_3O_4 from aggregation. The SSA of the composite increased from 4 to 47 m^2/g due to the introduction of ~3.3% graphene, which was favorable for electrolyte access. Thus, the Fe_3O_4/graphene composite displayed high rate performance and cycling stability due to the small size of Fe_3O_4 along with the improved conductivity and electrolyte accessibility. The composite could achieve a high capacity over 1000 mAh/g at 90 mA/g and good rate performance of 704 mAh/g at 2700 mA/g. The composite retained 90% of its capacity after 1000 cycles, and the Coulombic efficiency was nearly 100% in each cycle. The LISC based on the Fe_3O_4/graphene anode exhibited energy densities of 204–65 W h/kg over power densities from 55 to 4600 W/kg.

The theoretical capacity of Fe_2O_3 is 1005 mAh/g. However, its low conductivity and strong aggregation limit its application in LISCs. The intercalation of lithium-ions into Fe_2O_3 takes place between 0.2 and 1.5 V, while the deintercalation takes place between 1.8 and 2.5 V. Anchoring Fe_2O_3 particles in conducting carbon matrices can provide high conductivity and prevent Fe_2O_3 particles from aggregation during charge/discharge as well. To avoid the conventional slurry-processed electrodes using a polymeric binder, Zhao et al. fabricated a α-Fe_2O_3/MWCNT thin film based on hydrothermal method and spray deposition technique [73]. In the film, the MWCNTs formed a well-entangled and interconnected nanoporous structure, with the α-Fe_2O_3 spherical particles of 100 nm grown on the MWCNTs. A LISC was assembled with the α-Fe_2O_3/MWCNT thin film as anode and MWCNTs as cathode. The LISC could deliver a very high specific energy density of 50 W h/kg at a specific power density of 1000 W/kg over the potential range of 0–2.8 V. The superior performance of the LISC could be ascribed to the thin-film electrode without binders because the thickness and the binders might diminish the accessible specific area of the active materials, undermining effective ion transport and energy storage. In addition, MWCNTs greatly improved the conductivity of the thin-film electrode, and the pores in the MWCNTs network could act as reservoir for lithium-ions, which was favorable for lithium-ion transportation.

2.5 Nb_2O_5

Nb_2O_5 is another promising candidate for the anode materials of LISC. Nb_2O_5 has a theoretical capacity of 200 mAh/g, and it can deliver a high power through a mainly pseudocapacitive reaction of lithium-ions on the (near) surface of the electrode. Nb_2O_5 has several crystal structures, and orthorhombic phase has shown the highest specific capacity compared with other phases such as pseudohexagonal and amorphous phase. However, the application of Nb_2O_5 anodes in LISCs has been restricted by the low electrical conductivity and difficult to control optimum crystal structure. Lim et al. synthesized a mesoporous Nb_2O_5/carbon nanocomposite via a one-pot method by using niobium precursors and an amphiphilic block copolymer [74]. The block polymer not only provided amorphous carbon coating but also prevented nano-sized Nb_2O_5 from aggregation. The Nb_2O_5 nanoparticles in the composite provided a shortened diffusion length and the large electrode/electrolyte interface area. The mesopores of the composite enhanced ionic transportation, while the uniform amorphous carbon improved electrical conductivity. Therefore, the mesoporous Nb_2O_5/carbon nanocomposite exhibited good rate capability and cycling stability. The LISC composed by the composite anode and AC cathode exhibited excellent energy and power densities (74 and 18,510 W/kg), with advanced cycle life (capacity retention: ~90% at 1000 mA/g after 1000 cycles) within potential range from 1.0 to 3.5 V. Wang et al. reported a facile preparation method for orthorhombic phase Nb_2O_5 nanowire structure with ultra-thin carbon coating by carbonization of polydopamine-coated Nb_2O_5 nanowires [75], and the carbon-coated Nb_2O_5 nanowire could also be used as the battery-type anode in LISCs.

2.6 SILICA

Si is a promising anode material candidate for hybrid supercapacitors because of its low working voltage (<0.5 V) and high specific capacity (>3500 mAh/g). However, Si suffers from fast capacity fading caused by its large volume change during lithium-ion intercalation/deintercalation, which might be solved by nano-sizing and carbon coating. Ran et al. prepared high-performance Si battery-type anodes by size reduction (ball milling), carbon coating (thermal decomposition of acetylene gas), and B-doping (pyrolysis of the mixture of SiO powder and B_2O_3) of commercial silicon monoxide [76]. During the B-doping, nano-meter-scaled crystalline Si precipitated, and B did not form an alloy with Si; instead, it doped into Si. The B-doped Si nanocrystals embedded in a SiO_2 matrix, which acted as a buffer layer for volume change of Si during lithiation/delithiation. The size reduction leads to short charge transport paths and thus increased material utilization, and B doping decreased charge transfer resistance. The obtained B-Si/SiO_2/C exhibited good cycling stability and high specific capacity and rate performance, delivering a capacity of 685 mAh/g at a current density of 6.4 A/g. A LISC was constructed by coupling the B-Si/SiO_2/C anode with porous spherical carbon (PSC) cathode. The mass ratio of PSC to B-Si/SiO_2/C is set at 2:1 to leave a margin of anode for gradual decomposition of Si due to structural degradation to obtain long cycling life. The LISC worked at a voltage window from 2.0 to 4.5 V, exhibiting a high energy density of 128 W h/kg at 1229 W/kg and an energy density of 89 W h/kg at 9704 W/kg (conventional supercapacitor). Moreover, the LISC retained 70% of its initial capacity after 6000 cycles. Konno et al. reported the synthesis of Si–C–O glass-like compound and the usage of this compound as anode in the assembly of a LISC [77]. Although, the compound contained 55% Si, its electrochemical behavior was similar to that of HC. After pre-lithiation by metallic lithium, the potential of the compound decreased to 0.5 V, which revealed the possible application of the Si–C–O compound as anode materials in LISCs.

2.7 SUMMARY

Although great effort has been put on the research of battery-type anode for LISCs, none of the above anode materials such as carbon, titanium oxides, vanadium compounds, iron oxides, niobium(V) oxide, and silica can satisfy all the four requirements of an ideal anode material. There is still a long way to go before the battery-type anode materials of high electrochemical performance are discovered.

REFERENCES

1. Wu YP, Rahm E, Holze R. Carbon anode materials for lithium ion batteries. *Journal of Power Sources*. 2003;114(2):228–36.
2. Aravindan V, Gnanaraj J, Lee Y-S, Madhavi S. Insertion-type electrodes for nonaqueous Li-ion capacitors. *Chemical Reviews*. 2014;114(23):11619–35.
3. Zhang J, Shi Z, Wang J, Shi J. Composite of mesocarbon microbeads/hard carbon as anode material for lithium ion capacitor with high electrochemical performance. *Journal of Electroanalytical Chemistry*. 2015;747:20–8.
4. Kim H, Park K-Y, Cho M-Y, Kim M-H, Hong J, Jung S-K, et al. High-performance hybrid supercapacitor based on graphene-wrapped $Li_4Ti_5O_{12}$ and activated carbon. *ChemElectroChem*. 2014;1(1):125–30.
5. Khomenko V, Raymundo-Piñero E, Béguin F. High-energy density graphite/AC capacitor in organic electrolyte. *Journal of Power Sources*. 2008;177(2):643–51.
6. Billaud D, Henry FX, Willmann P. Electrochemical synthesis of binary graphite-lithium intercalation compounds. *Materials Research Bulletin*. 1993;28(5):477–83.
7. Ohzuku T, Iwakoshi Y, Sawai K. Formation of lithium-graphite intercalation compounds in nonaqueous electrolytes and their application as a negative electrode for a lithium ion (shuttlecock) cell. *Journal of the Electrochemical Society*. 1993;140(9):2490–8.
8. Shim J, Striebel KA. The dependence of natural graphite anode performance on electrode density. *Journal of Power Sources*. 2004;130(1):247–53.
9. Sole C, Drewett NE, Hardwick LJ. In situ Raman study of lithium-ion intercalation into microcrystalline graphite. *Faraday Discussions*. 2014;172:223–37. http://pubs.rsc.org/en/content/articlehtml/2014/fd/c4fd00079j.
10. Zaghib K, Nadeau G, Kinoshita K. Effect of graphite particle size on irreversible capacity loss. *Journal of the Electrochemical Society*. 2000;147(6):2110–5.
11. Sivakkumar SR, Pandolfo AG. Evaluation of lithium-ion capacitors assembled with pre-lithiated graphite anode and activated carbon cathode. *Electrochimica Acta*. 2012;65:280–7.
12. Decaux C, Lota G, Raymundo-Piñero E, Frackowiak E, Béguin F. Electrochemical performance of a hybrid lithium-ion capacitor with a graphite anode preloaded from lithium bis(trifluoromethane)sulfonimide-based electrolyte. *Electrochimica Acta*. 2012;86:282–6.
13. Saito M, Yamada T, Yodoya C, Kamei A, Hirota M, Takenaka T, et al. Influence of Li diffusion distance on the negative electrode properties of Si thin flakes for Li secondary batteries. *Solid State Ionics*. 2012;225:506–9.
14. Kim M, Xu F, Lee JH, Jung C, Hong SM, Zhang QM, et al. A fast and efficient pre-doping approach to high energy density lithium-ion hybrid capacitors. *Journal of Materials Chemistry A*. 2014;2(26):10029–33.
15. Ahn W, Lee DU, Li G, Feng K, Wang X, Yu A, et al. Highly oriented graphene sponge electrode for ultra high energy density lithium ion hybrid capacitors. *ACS Applied Materials & Interfaces*. 2016;8(38):25297–305.

16. Park H, Kim M, Xu F, Jung C, Hong SM, Koo CM. In situ synchrotron wide-angle X-ray scattering study on rapid lithiation of graphite anode via direct contact method for Li-ion capacitors. *Journal of Power Sources.* 2015;283:68–73.
17. Park M-S, Lim Y-G, Kim J-H, Kim Y-J, Cho J, Kim J-S. A Novel lithium-doping approach for an advanced lithium ion capacitor. *Advanced Energy Materials.* 2011;1(6):1002–6.
18. Kobayashi H, Tabuchi M, Shikano M, Kageyama H, Kanno R. Structure, and magnetic and electrochemical properties of layered oxides, Li_2IrO_3. *Journal of Materials Chemistry.* 2003;13(4):957–62.
19. Sivakkumar SR, Milev AS, Pandolfo AG. Effect of ball-milling on the rate and cycle-life performance of graphite as negative electrodes in lithium-ion capacitors. *Electrochimica Acta.* 2011;56(27):9700–6.
20. Lee JH, Shin WH, Lim SY, Kim BG, Choi JW. Modified graphite and graphene electrodes for high-performance lithium ion hybrid capacitors. *Materials for Renewable and Sustainable Energy.* 2014;3(1):22.
21. Kim J-H, Kim J-S, Lim Y-G, Lee J-G, Kim Y-J. Effect of carbon types on the electrochemical properties of negative electrodes for Li-ion capacitors. *Journal of Power Sources.* 2011;196(23):10490–5.
22. Ni J, Huang Y, Gao L. A high-performance hard carbon for Li-ion batteries and supercapacitors application. *Journal of Power Sources.* 2013;223:306–11.
23. Liu Y, Xue JS, Zheng T, Dahn JR. Mechanism of lithium insertion in hard carbons prepared by pyrolysis of epoxy resins. *Carbon.* 1996;34(2):193–200.
24. Zhang J, Liu X, Wang J, Shi J, Shi Z. Different types of pre-lithiated hard carbon as negative electrode material for lithium-ion capacitors. *Electrochimica Acta.* 2016;187:134–42.
25. Cao WJ, Zheng JP. Li-ion capacitors with carbon cathode and hard carbon/stabilized lithium metal powder anode electrodes. *Journal of Power Sources.* 2012;213:180–5.
26. Wang G, Shen X, Yao J, Park J. Graphene nanosheets for enhanced lithium storage in lithium ion batteries. *Carbon.* 2009;47(8):2049–53.
27. Ren JJ, Su LW, Qin X, Yang M, Wei JP, Zhou Z, et al. Pre-lithiated graphene nanosheets as negative electrode materials for Li-ion capacitors with high power and energy density. *Journal of Power Sources.* 2014;264:108–13.
28. Acznik I, Lota K, Sierczyńska A. Improvement of power–energy characteristic of the lithium-ion capacitor by structure modification of the graphite anode. *MRS Communications.* 2017;7(2):245–52.
29. Sun Y, Tang J, Qin F, Yuan J, Zhang K, Li J, et al. Hybrid lithium-ion capacitors with asymmetric graphene electrodes. *Journal of Materials Chemistry A.* 2017;5(26):13601–9.
30. Zhang J, Shi Z, Wang C. Effect of pre-lithiation degrees of mesocarbon microbeads anode on the electrochemical performance of lithium-ion capacitors. *Electrochimica Acta.* 2014;125:22–8.
31. Schroeder M, Winter M, Passerini S, Balducci A. On the cycling stability of lithium-ion capacitors containing soft carbon as anodic material. *Journal of Power Sources.* 2013;238:388–94.
32. Yang Z, Choi D, Kerisit S, Rosso KM, Wang D, Zhang J, et al. Nanostructures and lithium electrochemical reactivity of lithium titanites and titanium oxides: A review. *Journal of Power Sources.* 2009;192(2):588–98.
33. Noailles LD, Johnson CS, Vaughey JT, Thackeray MM. Lithium insertion into hollandite-type TiO_2. *Journal of Power Sources.* 1999;81:259–63.
34. Song T, Han H, Choi H, Lee JW, Park H, Lee S, et al. TiO_2 nanotube branched tree on a carbon nanofiber nanostructure as an anode for high energy and power lithium ion batteries. *Nano Research.* 2014;7(4):491–501.

35. Cho M-Y, Park S-M, Kim K-B, Roh KC. Synthesis of mesoporous spherical TiO_2 and its application in negative electrode of hybrid supercapacitor. *Electronic Materials Letters*. 2013;9(6):809–12.
36. Cai Y, Zhao B, Wang J, Shao Z. Non-aqueous hybrid supercapacitors fabricated with mesoporous TiO_2 microspheres and activated carbon electrodes with superior performance. *Journal of Power Sources*. 2014;253:80–9.
37. Wang H, Guan C, Wang X, Fan HJ. A high energy and power Li-ion capacitor based on a TiO_2 nanobelt array anode and a graphene hydrogel cathode. *Small*. 2015;11(12):1470–7.
38. Kim JH, Choi HJ, Kim H-K, Lee S-H, Lee Y-H. A hybrid supercapacitor fabricated with an activated carbon as cathode and an urchin-like TiO_2 as anode. *International Journal of Hydrogen Energy*. 2016;41(31):13549–56.
39. Choi H-J, Kim JH, Kim H-K, Lee S-H, Lee Y-H. Improving the electrochemical performance of hybrid supercapacitor using well-organized urchin-like TiO_2 and activated carbon. *Electrochimica Acta*. 2016;208:202–10.
40. Jianli C, Guifang G, Wei N, Qun G, Yinchuan L, Bin W. Graphene oxide hydrogel as a restricted-area nanoreactor for synthesis of 3D graphene-supported ultrafine TiO_2 nanorod nanocomposites for high-rate lithium-ion battery anodes. *Nanotechnology*. 2017;28(30):305401.
41. Chen Z, Yuan Y, Zhou H, Wang X, Gan Z, Wang F, et al. 3D nanocomposite architectures from carbon-nanotube-threaded nanocrystals for high-performance electrochemical energy storage. *Advanced Materials*. 2014;26(2):339–45.
42. Gao L, Huang D, Shen Y, Wang M. Rutile-TiO_2 decorated $Li_4Ti_5O_{12}$ nanosheet arrays with 3D interconnected architecture as anodes for high performance hybrid supercapacitors. *Journal of Materials Chemistry A*. 2015;3(46):23570–6.
43. Dylla AG, Henkelman G, Stevenson KJ. Lithium insertion in nanostructured TiO_2(B) architectures. *Accounts of Chemical Research*. 2013;46(5):1104–12.
44. Dylla AG, Xiao P, Henkelman G, Stevenson KJ. Morphological dependence of lithium insertion in nanocrystalline TiO_2(B) nanoparticles and nanosheets. *The Journal of Physical Chemistry Letters*. 2012;3(15):2015–9.
45. Aravindan V, Shubha N, Ling WC, Madhavi S. Constructing high energy density non-aqueous Li-ion capacitors using monoclinic TiO_2-B nanorods as insertion host. *Journal of Materials Chemistry A*. 2013;1(20):6145–51.
46. Wang Q, Wen ZH, Li JH. A hybrid supercapacitor fabricated with a carbon nanotube cathode and a TiO_2–B nanowire anode. *Advanced Functional Materials*. 2006;16(16):2141–6.
47. Naoi K, Kurita T, Abe M, Furuhashi T, Abe Y, Okazaki K et al. Ultrafast nanocrystalline-TiO_2(B)/carbon nanotube hyperdispersion prepared via combined ultracentrifugation and hydrothermal treatments for hybrid supercapacitors. *Advanced Materials*. 2016;28(31):6751–7.
48. Liu H, Bi Z, Sun X-G, Unocic RR, Paranthaman MP, Dai S et al. Mesoporous TiO_2–B microspheres with superior rate performance for lithium ion batteries. *Advanced Materials*. 2011;23(30):3450–4.
49. Shu J. Electrochemical behavior and stability of $Li_4Ti_5O_{12}$ in a broad voltage window. *Journal of Solid State Electrochemistry*. 2009;13(10):1535–9.
50. Yi T-F, Yang S-Y, Xie Y. Recent advances of $Li_4Ti_5O_{12}$ as a promising next generation anode material for high power lithium-ion batteries. *Journal of Materials Chemistry A*. 2015;3(11):5750–77.
51. Lu X, Zhao L, He X, Xiao R, Gu L, Hu Y-S et al. Lithium storage in $Li_4Ti_5O_{12}$ spinel: The full static picture from electron microscopy. *Advanced Materials*. 2012;24(24):3233–8.
52. Jansen AN, Kahaian AJ, Kepler KD, Nelson PA, Amine K, Dees DW et al. Development of a high-power lithium-ion battery. *Journal of Power Sources*. 1999;81:902–5.

53. Naoi K, Ishimoto S, Isobe Y, Aoyagi S. High-rate nano-crystalline $Li_4Ti_5O_{12}$ attached on carbon nano-fibers for hybrid supercapacitors. *Journal of Power Sources.* 2010;195(18):6250–4.
54. Guerfi A, Sévigny S, Lagacé M, Hovington P, Kinoshita K, Zaghib K. Nano-particle $Li_4Ti_5O_{12}$ spinel as electrode for electrochemical generators. *Journal of Power Sources.* 2003;119:88–94.
55. Cheng L, Liu H-J, Zhang J-J, Xiong H-M, Xia Y-Y. Nanosized $Li_4Ti_5O_{12}$ prepared by molten salt method as an electrode material for hybrid electrochemical supercapacitors. *Journal of the Electrochemical Society.* 2006;153(8):A1472–A7.
56. Lee B-G, Lee S-H. Application of hybrid supercapacitor using granule $Li_4Ti_5O_{12}$/activated carbon with variation of current density. *Journal of Power Sources.* 2017;343:545–9.
57. Choi HS, Im JH, Kim T, Park JH, Park CR. Advanced energy storage device: A hybrid BatCap system consisting of battery-supercapacitor hybrid electrodes based on $Li_4Ti_5O_{12}$-activated-carbon hybrid nanotubes. *Journal of Materials Chemistry.* 2012;22(33):16986–93.
58. Hong Soo C, TaeHoon K, Ji Hyuk I, Chong Rae P. Preparation and electrochemical performance of hyper-networked $Li_4Ti_5O_{12}$/carbon hybrid nanofiber sheets for a battery–supercapacitor hybrid system. *Nanotechnology.* 2011;22(40):405402.
59. Zuo W, Wang C, Li Y, Liu J. Directly grown nanostructured electrodes for high volumetric energy density binder-free hybrid supercapacitors: A case study of CNTs//$Li_4Ti_5O_{12}$. 2015;5:7780.
60. Patoux S, Masquelier C. Lithium insertion into titanium phosphates, silicates, and sulfates. *Chemistry of Materials.* 2002;14(12):5057–68.
61. Kuhn A, Martín M, García-Alvarado F. Synthesis, structure and electrochemical lithium intercalation chemistry of ramsdellite-type $LiCrTiO_4$. *Zeitschrift für anorganische und allgemeine Chemie.* 2008;634(5):880–6.
62. Liu D, Ouyang C, Shu J, Jiang J, Wang Z, Chen L. Theoretical study of cation doping effect on the electronic conductivity of $Li_4Ti_5O_{12}$. *physica status solidi (b).* 2006;243(8):1835–41.
63. Mukai K, Ariyoshi K, Ohzuku T. Comparative study of $Li[CrTi]O_4$, $Li[Li_{1/3}Ti_{5/3}]O_4$ and $Li_{1/2}Fe_{1/2}[Li_{1/2}Fe_{1/2}Ti]O_4$ in non-aqueous lithium cells. *Journal of Power Sources.* 2005;146(1):213–6.
64. Rao CV, Rambabu B. Nanocrystalline $LiCrTiO_4$ as anode for asymmetric hybrid supercapacitor. *Solid State Ionics.* 2010;181(17):839–43.
65. Aravindan V, Chuiling W, Madhavi S. High power lithium-ion hybrid electrochemical capacitors using spinel $LiCrTiO_4$ as insertion electrode. *Journal of Materials Chemistry.* 2012;22(31):16026–31.
66. Aatiq A, Menetrier M, Croguennec L, Suard E, Delmas C. On the structure of $Li_3Ti_2(PO_4)_3$. *Journal of Materials Chemistry.* 2002;12(10):2971–8.
67. Luo J-Y, Xia Y-Y. Electrochemical profile of an asymmetric supercapacitor using carbon-coated $LiTi_2(PO_4)_3$ and active carbon electrodes. *Journal of Power Sources.* 2009;186(1):224–7.
68. Aravindan V, Chuiling W, Reddy MV, Rao GVS, Chowdari BVR, Madhavi S. Carbon coated nano-$LiTi_2(PO_4)_3$ electrodes for non-aqueous hybrid supercapacitors. *Physical Chemistry Chemical Physics.* 2012;14(16):5808–14.
69. Shi Z, Wang Q, Ye W, Li Y, Yang Y. Synthesis and characterization of mesoporous titanium pyrophosphate as lithium intercalation electrode materials. *Microporous and Mesoporous Materials.* 2006;88(1):232–7.
70. Whittingham MS, Song Y, Lutta S, Zavalij PY, Chernova NA. Some transition metal (oxy)phosphates and vanadium oxides for lithium batteries. *Journal of Materials Chemistry.* 2005;15(33):3362–79.

71. Satish R, Aravindan V, Ling WC, Madhavi S. Carbon-coated $Li_3V_2(PO_4)_3$ as insertion type electrode for lithium-ion hybrid electrochemical capacitors: An evaluation of anode and cathodic performance. *Journal of Power Sources*. 2015;281:310–7.
72. Zhang F, Zhang T, Yang X, Zhang L, Leng K, Huang Y et al. A high-performance supercapacitor-battery hybrid energy storage device based on graphene-enhanced electrode materials with ultrahigh energy density. *Energy & Environmental Science*. 2013;6(5):1623–32.
73. Zhao X, Johnston C, Grant PS. A novel hybrid supercapacitor with a carbon nanotube cathode and an iron oxide/carbon nanotube composite anode. *Journal of Materials Chemistry*. 2009;19(46):8755–60.
74. Lim E, Kim H, Jo C, Chun J, Ku K, Kim S et al. Advanced hybrid supercapacitor based on a mesoporous niobium pentoxide/carbon as high-performance anode. *ACS Nano*. 2014;8(9):8968–78.
75. Wang X, Yan C, Yan J, Sumboja A, Lee PS. Orthorhombic niobium oxide nanowires for next generation hybrid supercapacitor device. *Nano Energy*. 2015;11:765–72.
76. Yi R, Chen S, Song J, Gordin ML, Manivannan A, Wang D. High-performance hybrid supercapacitor enabled by a high-rate Si-based anode. *Advanced Functional Materials*. 2014;24(47):7433–9.
77. Konno H, Kasashima T, Azumi K. Application of Si–C–O glass-like compounds as negative electrode materials for lithium hybrid capacitors. *Journal of Power Sources*. 2009;191(2):623–7.

3 Cathodes of Lithium-Ion Supercapacitors

Kun Feng and Zhongwei Chen
University of Waterloo

CONTENTS

3.1 CONFIGURATION AND CURRENT STATUS OF LITHIUM-ION SUPERCAPACITORS BASED ON ACTIVATED CARBON CATHODES

The most popular and only commercialized cathode material in Lithium-Ion Supercapacitors (LISCs) is activated carbon (AC), which is also the widely used electrode material in traditional electrochemical double-layer capacitors (EDLCs). A schematic presentation of a typical LISC is shown in Figure 3.1. PF_6^- anodes migrate to the cathode side of the LISC and are adsorbed on the surface of AC during charging process, transferring electrical energy into chemical energy. The prevalence of AC in supercapacitors is mainly due to its large surface area and porous structure, which can host large quantities of charged species and provide fast charge transfer. Other merits of AC including easy accessibility, chemical inertness, and low cost further contribute to the extensive application of AC in SCs [1].

The first commonly acknowledged LISC was reported, named, and patented by Amatucci et al. [3]. In their system, $Li_4Ti_5O_{12}$ (LTO) was used as an intercalation-type anode; AC was used as a capacitive cathode; and lithium-ion battery (LIB) electrolyte with composition of 1 M $LiPF_6$ in the 2:1 volume ratio of ethylene carbonate (EC):dimethyl carbonate (DMC) was used as an electrolyte for the LISC. A sloping

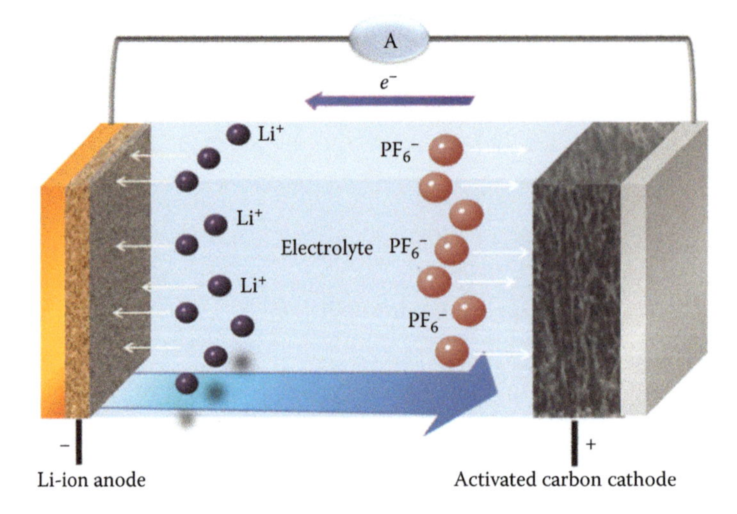

FIGURE 3.1 Charging process of a hybrid supercapacitor using AC as the cathode and a Li-insertion material as the anode. Reprinted with permission from Wiley [2].

voltage profile in the voltage range of 1.5–3 V was observed, and 90% capacity retention was achieved when the charge/discharge rate was 10 C (Figure 3.2). After 5,000 cycles, 85%–90% of initial capacity was retained. Packaged energy densities were calculated to be more than 20 W h kg^{-1} in their flat plate plastic cell, which was 400%–500% increase with respect to EDLCs in the same era.

With AC functioning as capacitive cathode, a wide spectrum of anode materials has been paired with AC and formed LISCs with different properties and performances. Some nanostructured battery-type anode materials and lithium-containing compounds have been discovered to have the capability of fast charge and discharge, which qualifies them as anode materials in AC-based LISCs. These materials include

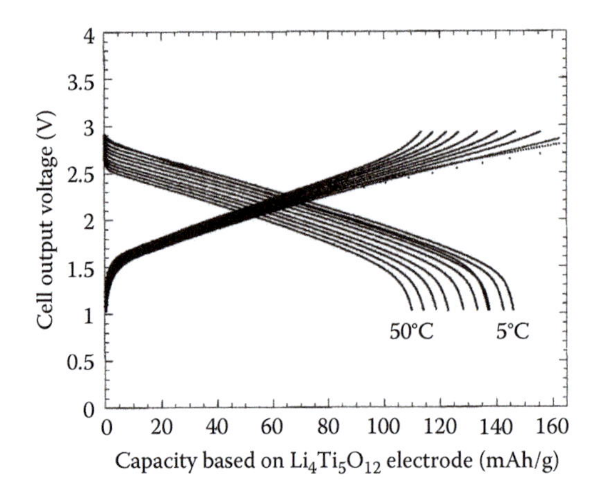

FIGURE 3.2 Discharge profile as a function of rate for plastic AC/LTO cell. Specific capacity of LTO is shown. Reprinted with permission from the Electrochemical Society [3].

nano-structured LTO [4], lithium silicates, [5] N_2O_5, [6] TiO_2-B, [7] etc. [8]. Prelithiated carbon-based anodes coupling with AC cathodes can provide a cell voltage above 3.8 V. These materials include different types of graphite [9,10], nongraphitizable carbon [11]. Zhang et al. reported a remarkable energy density of 92.3 W h kg^{-1} and power density of 5.5 kW kg^{-1}, achieved by the combination of prelithiated meso-carbon microbeads and commercial AC [12]. Good availability of both the cathode and the anode materials, along with improved performance of carbon-based LISCs, have presented a promising future for this system. It should be noted that this type of anode usually suffers from lithium dendrite formation and SEI growth at the anode–electrolyte interface due to the low intercalation potentials of these materials. These problems may lead to an increase in resistance and hinder device performance. Severe dendrite formations can pierce separators and cause potential safety hazards in some extreme cases. Another challenge with this type of LISCs is the prelithiation of anode materials. Carbon-based anodes do not bear Li ions and thus require a prelithiation step, which is critical for the performance of anodes, and could be technically challenging on an industrial scale. Despite the challenges revealed, this type of LISC has gained much attention from industry, with many companies providing commercial products based on the same criteria. Some commercialized products from several major Japanese companies are listed in Table 3.1, with specs of each product listed. Reprinted with permission from the Electrochemical Society [13].

3.2 DEVELOPMENT OF ACTIVATED CARBON CATHODES

On the one hand, it is encouraging that commercial AC can be directly applied in the fabrication of LISCs. On the other hand, it is desirable and worthwhile to put efforts into the modification and improvement of AC, in terms of both fabrication and material itself. Alternative fabrication procedures and carbon sources can further reduce the cost and simplify the process for large-scale production. A controllable tuning on physical and chemical characteristics such as microstructure (pore size, surface

TABLE 3.1
Commercialized LISC Devices from Five Major Japanese Companies in 2008–2011

Company	Cell Type	Cell Size (mm)	Rated Voltage (V)	Capacitance (F)	Energy Density (W h kg^{-1})
JM energy	Laminate	180 × 126 × 10.9	2.2–3.8	2,000	14
Asahi Kasei FDK energy	Module (laminate)	106 × 138 × 8.5	15	600	12 W h (module)
TAIYO YUDEN	Cylinder	25 ∅ × 40 L	2.2–3.8	200	9.8
ACT	Laminate	100 × 100 × 20.1	2.0–4.0	5,000	15
NEC Tokin	Laminate	192 × 95 × 5.0	2.2–3.8	1,000	8.0

Source: Reprinted with permission from the Royal Society of Chemistry [13].

area), functional group, and heteroatom doping can help obtain desired performance enhancement in some specific respects, including cycle life, specific capacity, and rate performance.

Several research groups have reported successful conversion of AC from biomass materials. Jain et al. carbonized coconut shell under different pretreatment, carbonization, and oxidation conditions, and obtained high surface area and mesoporous AC. They achieved a dramatic increase in energy density (69 W h kg^{-1}) with the synthesized AC and commercialized LTO [14]. Sun et al. prepared microporous AC with high surface area from fresh pomelo peel [15]. The biomass was precarbonized in N_2 and activated in a KOH solution. An alloy type Sn–C anode was coupled with the synthesized AC in their LISC system, which delivered energy densities of 195.7 and 84.6 W h kg^{-1}, when the system was tested at 731.25 and 24,375 W kg^{-1}, respectively. Li et al. developed a type of AC containing rich oxygen functional groups from biomass transfer of egg white, with high surface areas and narrow pore-size distribution in the range of 0.5–2 nm [16]. A full-cell LISC was assembled with the as-prepared AC as the cathode, the Si/C nano composite as the anode, and a traditional LIB electrolyte. The electrochemical performance of the LISC is shown in Figure 3.3. Figure 3.3a displays the voltage profiles of the LISC with AC heat-treated at 900°C. Current densities used

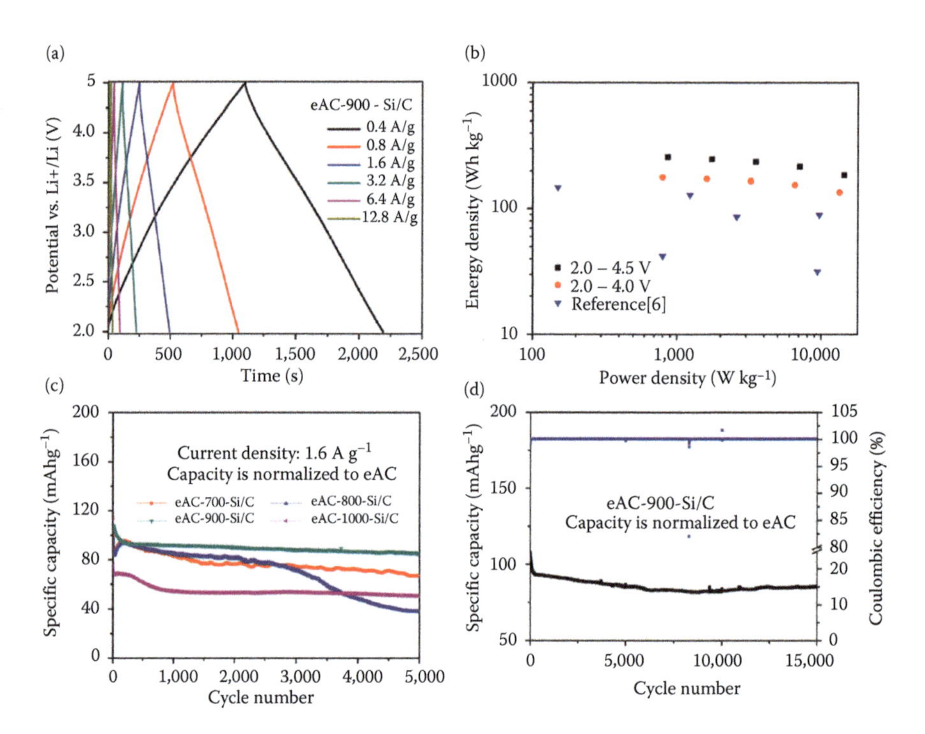

FIGURE 3.3 (a) Voltage profile of Si/C//eAC-900 LISC full cell using galvanostatic charge/discharge; (b) Ragone plot of this work in comparison with energy densities at maximum power densities of previous publication [6]; (c) cycling performance of Si/C//eAC-700, Si/C//eAC-800, Si/C//eAC-900, and Si/C//eAC-1000 LISC full cells at 1.6 A g^{-1}; and (d) cycling performance of Si/C//eAC-900 LISC full cell at 1.6 A g^{-1}. Reprinted with permission from Wiley [16].

in the galvanostatic charge/discharge range from 0.4 to 12.8 A g^{-1}. The voltage curves show near-linear shapes, resembling an EDLC. With the increase of current density, more battery characteristics appear. During the charge process, Li-ion alloy with the Si/C anode, which follows a faradaic process. On the cathode side, PF_6^- anions adsorb on the surface of AC, following a nonfaradaic process. The charge and discharge processes are highly reversible as displayed by the voltage curves. The Ragone plot in Figure 3.3b compares the LISC system with a previously reported LISC with similar electrode materials and testing conditions. Based on the total mass of active materials on both the anode and the cathode, their LISC exhibits a high energy density of 258 W h kg^{-1} at the power density of 867 W kg^{-1}. A remarkable energy density of 147 W h kg^{-1} is retained when the power density is boosted to 29,893 W kg^{-1}, outperforming most LISCs in recent publications. ACs prepared under different conditions are tested, and their cycling performances are displayed in Figure 3.3c. LISCs prepared with eAC-900-Si/C and eAC-1000-Si/C show more stable cycling than the other two LISCs. Figure 3.3d shows the long-term cycling performance of a LISC, with a voltage window of 2.0–4.0 V, a capacity retention of 86%, and a near-100% coulombic efficiency. The decrease of operation voltage plays a significant role in the improvement of cycling stability, despite a 30% decrease of energy density, initiating a debate on the trade-off between energy density and cycling stability.

On par with the modification of production procedure and original source of AC, research efforts on the improvement AC have made much progress in enhancing the performance of AC. Despite the high surface area and high porosity of commercially available AC, its moderate electronic conductivity and nonwettable micropores may prevent fast migration of charged species and diffusion of electrolyte, restricting the full capacitance exertion of AC. Thus, research efforts on regulating the structure and surface properties of AC bear significant importance in tackling the aforementioned problems and providing improved performance [17,18].

Kim et al. reported an electroless deposition method, where Cu nanoparticles were deposited on AC [19]. The deposited Cu NPs contribute to a significant enhancement in electronic conductivity of AC and improve both its specific capacitance and rate performance. At a charge/discharge rate of 100 C, the Cu-deposited AC electrode shows an 11% increase in capacitance over the untreated AC. Liu et al. discovered the functionalization of commercial powder-form AC can be effective on the improvement of its capacitance due to the pseudocapacitance induced by the functional groups on AC surface [18]. After a solution treatment of AC, during which AC was added to a mixed solution of 1.0 M $(NH_4)_2S_2O_8$ and 2.0 M H_2SO_4, and stirred at 60 °C for 6 h, its surface area was reduced to a quarter of its original value. Despite the decrease in surface area due to functionalization, capacitance more than tripled after the treatment. They attributed the increase of capacitance to the C=O functional groups, which could react with Li^+ reversibly in the voltage range of 1.5–3.5 V Li/Li^+ and generate pseudocapacitance.

Heteroatom doping on carbon materials is another popular approach in achieving desirable physical and chemical properties for AC [20]. It was suggested by some researchers that heteroatom doping can have a positive impact on the conductivity of carbon materials. In addition, wettability of electrolytes with pore walls can be improved by nitrogen-containing functional groups, which can also result in more capacitance from faradaic reaction [21]. Li et al. introduced nitrogen doping via an ammonia treatment process [22]. The pore size distribution of the obtained nitrogen-doped ACs (NACs) is narrow, with most pores in the category of meso- and micropores. Four percent of N content can be achieved, and the specific surface area can reach as high as 2,859 m^2 g^{-1}. As shown in Figure 3.4a, NAC treated at 400°C shows best cycling stability, with high initial specific capacity of 127 mAh g^{-1} (182 F g^{-1}), with 86% retention after 500 cycles. Quasi-rectangular cyclic voltammetry (CV) profiles in Figure 3.4b suggest capacitive behavior of all NACs. Rate performance of NACs is displayed in Figure 3.4c, with current densities ranging from 0.4 to 12.8 A g^{-1}. Specific capacity of 73.7 mAh g^{-1} is retained by NAC-400 at a high current density of 12.8 A g^{-1}, higher than the other two samples. Near-linear charge/discharge curves of NAC-400 are shown in Figure 3.4d at different current densities. The performance of NACs highly depends on the N content of these materials. N content of 2.97% in NAC-400 provides superior performance over NAC-0 and NAC-600, whose N contents are 0% and 3.98%, respectively. It should be noted that higher N content does not necessarily provide better performance, possibly due to much increased oxidized pyridinic N and reduced specific surface area in NAC-600,

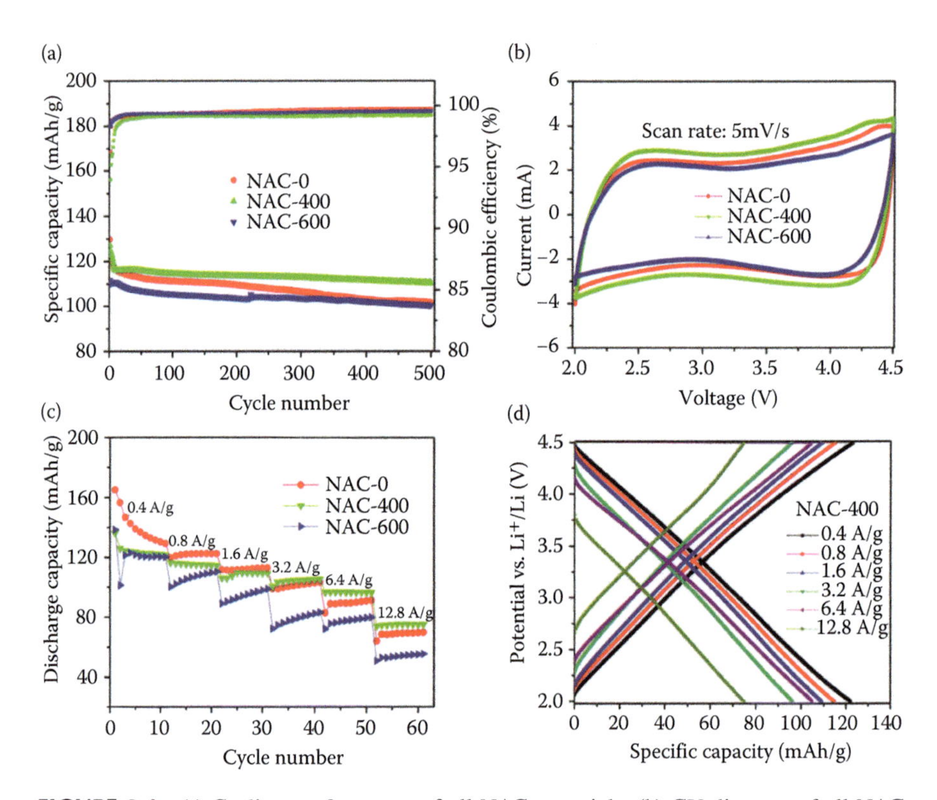

FIGURE 3.4 (a) Cycling performance of all NAC materials, (b) CV diagram of all NAC materials at a scan rate of 5 mV s^{-1}, (c) rate performance of all NAC materials, and (d) voltage profiles of NAC-400 at different current densities. Reprinted with permission from the Royal Society of Chemistry [22].

which can attenuate the ion adsorption power of NAC. Although the electronic conductivity can be increased by N doping, a careful balance between this improvement and the associated drawbacks should be considered to achieve the best performance. A full-cell LISC, using NAC-400 as the cathode and Si/C as the anode, was fabricated and tested. A high energy density of 230 W h kg^{-1} was achieved when the LISC was tested at 1,747 W kg^{-1}, and 141 W h kg^{-1} at a high power density of 30,127 W kg^{-1}, the performance of which was among the highest of all previously reported LISCs. The energy density of this device is on the same order of LIBs, while the power density can reach much higher than that of LIBs.

3.3 GRAPHENE-BASED CATHODES

Research on other carbon materials (graphene, carbon nanotubes [CNTs], etc.) is on the surge due to some unique properties of these materials. Graphene has garnered much attention in the field of electrochemical energy storage devices in past years. It has comparable theoretical specific surface area with AC. Meanwhile, the electronic conductivity of graphene is much higher than that of AC. In addition, graphene has excellent thermal stability and mechanical strength. These properties make it a good candidate as an electrode material in batteries and SCs [23]. In this chapter, reduced graphene oxide (rGO) is categorized as graphene for simplicity purpose, although technically graphene stands for an allotrope (form) of carbon consisting of single-layer thick hexagonal lattice with one atom at each vertex. CNTs are deemed ideal for batteries and SCs due to their advantages of continuous and interconnected conductive paths, good chemical and mechanical stability, and special pore structure [24]. Some other porous or hierarchical carbon materials have also been reported as cathode materials for LISCs with attractive performances [25,26].

Zhang et al. utilized a graphene-based three-dimensional porous carbon material (3D graphene) as the cathode, and Fe_3O_4/graphene (Fe_3O_4/G) nanocomposite as the anode for LISC [27]. The graphene precursor for both the cathode and the anode is graphene oxide (GO), which has good dispersion in water and many nonaqueous solvents. After the solvothermal/hydrothermal steps and annealing/activation procedures, both anode and cathode materials are obtained and ready to use without further treatment. The 3D graphene has a high SSA of 3,355 m^2 g^{-1}. Full-cell LISC with the synthesized materials displays a high energy density of 86 W h kg^{-1} at 2,587 W kg^{-1}.

A LISC with all graphene-based electrodes was later reported in 2015, in which graphene is the main contributor to the capacity and capacitance at anode and cathode sides [28]. Graphene used in this project is thin-layer and highly curved (Figure 3.5). The 3D graphene based electrode material has an ultra-high SSA of 3,523 m^2 g^{-1}, and narrowly distributed pore size (most of pores in the range of 1–7 nm), facilitating electrolyte accessibility. The conductivity of the bulk electrode materials is as high as 303 S m^{-1}, an order higher than that of AC. A symmetric SC was assembled with two identical 3D graphene electrodes. The capacitance of the cathode was found to be above 160 F g^{-1} at 0.1 A g^{-1}, and retained more than 80% when the charge/discharge rate was elevated to 5 A g^{-1}. The energy density of this symmetric SC was 41–27.7 W h g^{-1} in the current density range of 0.1–5 A g^{-1}. The anode is a freestanding graphene film prepared by vacuum filtration and subsequent

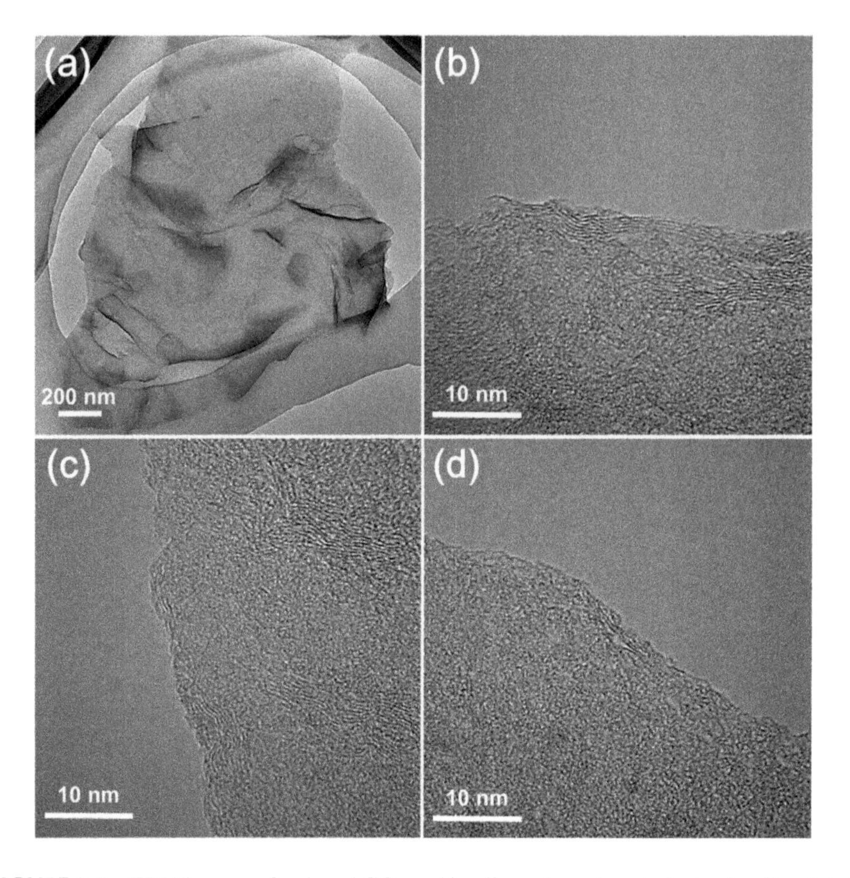

FIGURE 3.5 TEM images of reduced GO used in all graphene-based electrodes. Reprinted with permission from Elsevier [28].

flash-light reduction, using same GO source that was used in the cathode. The full-cell LISC can be operated to a high voltage of 4.2 V. It presented a high energy density of 71.5 W h kg^{-1}, when a high power density of 7,800 W kg^{-1} was applied.

Liu et al. developed 3D graphene-foam composites as LISC electrodes [29]. Three-dimensional molybdenum trioxide/graphene nanosheets (3D MoO_3/GNSs) foam was utilized as the anode, and 3D polyaniline/GNSs (3D PANI/GNSs) foam was adopted as the cathode. 3D PANI/GO hydrogel-like composite (Figure 3.6a), a precursor of 3D PANI/GNSs, was prepared via a hydrothermal reaction with GO and PANI as starting materials. After subsequent pyrolysis and activation, 3D PANI/GNSs composite was obtained, with SSA above 2,200 m^2 g^{-1}, and average pore size of 2.7 nm. SEM and TEM images (Figure 3.6b–e) show PANI-derived carbon tubes interweave with each other among the GNS matrix, with some tubes covered by GNS. Broad X-ray diffraction (XRD) peaks in Figure 3.6f show both the amorphous and graphitic nature of the composite.

Raman spectroscopy with D band and G band of carbon materials confirmed the disordered nature of carbon in this composite, consistent with the results of XRD and

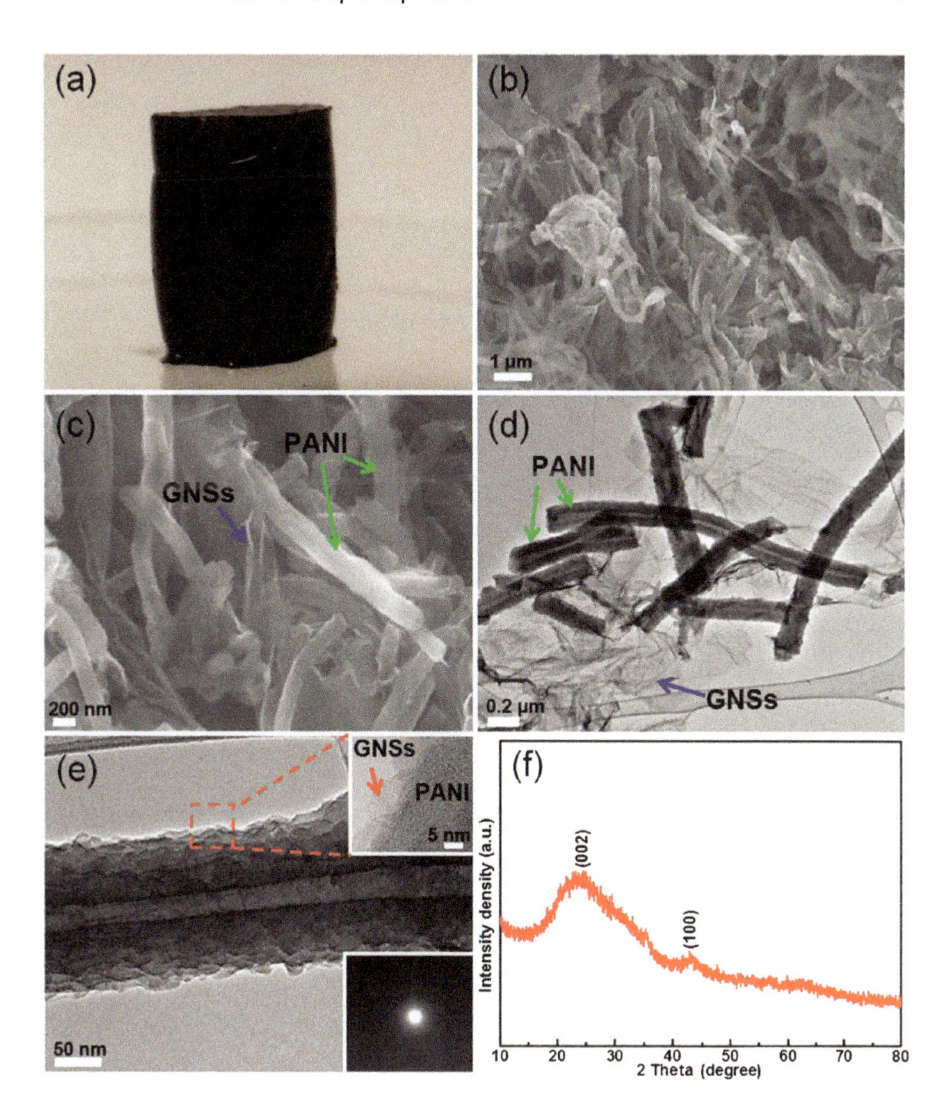

FIGURE 3.6 (a) Photograph, (b) low-magnification SEM image, (c) high-magnification SEM image, (d) low-resolution TEM image, (e) high-resolution TEM image, and (f) XRD spectrum of PANI/GNS-2-derived carbon. The inset shows the corresponding HRTEM image (top) and SAED pattern (down). Reprinted with permission from the American Chemical Society [29].

selected-area electron diffraction pattern (SAED) from TEM. PANI was used due to facile synthesis procedure, and good N-doping in carbon derived from it, which was confirmed by X-ray photoelectron spectroscopy (XPS) as shown in Figure 3.7. Both N and O functionalities were evidenced by XPS.

The all graphene-based LISC delivered a high energy density of 128.3 W h kg^{-1} at 182.2 W kg^{-1}. When a high power density of 13.5 kW kg^{-1} was used, 44.1 W h kg^{-1} of energy density was retained. Graphene in this LISC was claimed to be the main

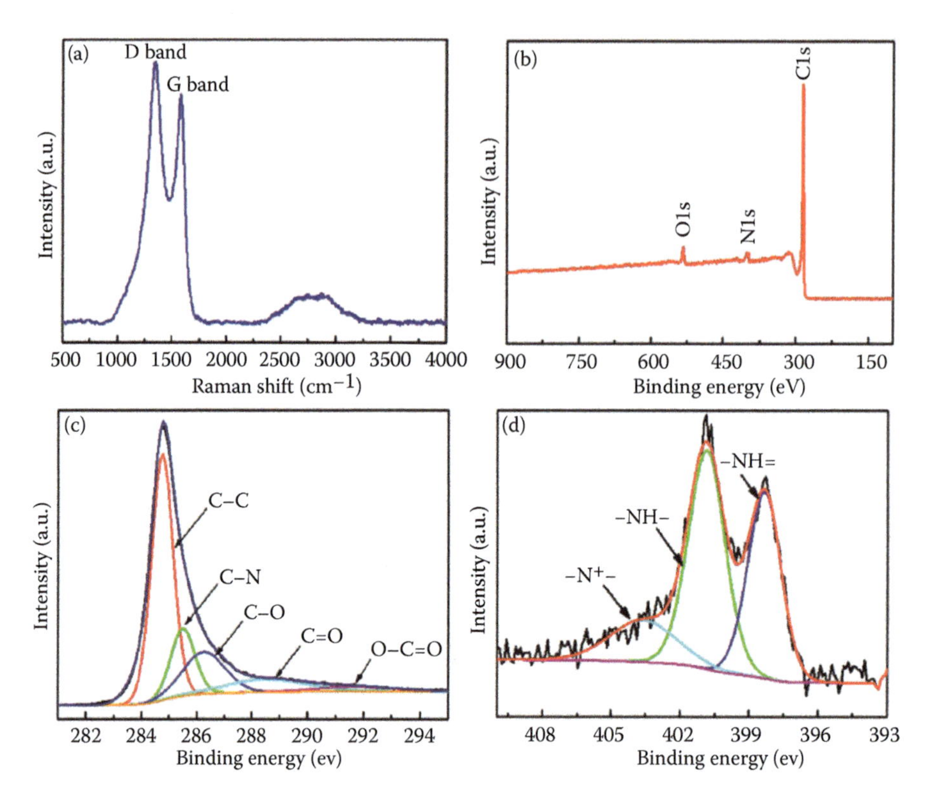

FIGURE 3.7 (a) Raman spectrum and XPS spectrum of the as-synthesized PANI/GNS-2 foam-derived carbon material: (b) survey, (c) C 1s, and (d) N 1s. Reprinted with permission from American Chemical Society [29].

contributor to the excellent performance, owing to its high surface area, good electronic conductivity, and the unique 2D structure which can act as support and host for other materials.

3.4 OTHER CARBON-BASED CATHODES

In 2006, Wang et al. first reported the application of CNTs in a LISC, seven years after the debut of LISCs [30]. Lab-synthesized multi-walled CNTs were utilized as the cathode and TiO_2-B nanowires were used as the anode. As one of the pioneering researchers employing CNTs as the cathode material in a LISC, authors reported a comparable energy density (12.5 W h kg^{-1}) to that of prevailing AC/LTO systems by then, which was encouraging and demonstrated the possibility of using CNTs as the LISC cathode. Zhao et al. used CNTs as both the cathode and the anode in their LISC device [31]. The cathode was MWCNT thin film, and the anode was α-Fe_2O_3/MWNT thin film, both of which were prepared by a spray deposition method. It should be noted that no additional conductive agent was used, and both electrodes were binderless. The electrodes possessed good flexibility as shown in Figure 3.8 and remained intact and bendable after cycling. Good electronic conductivity and excellent mechanical properties of

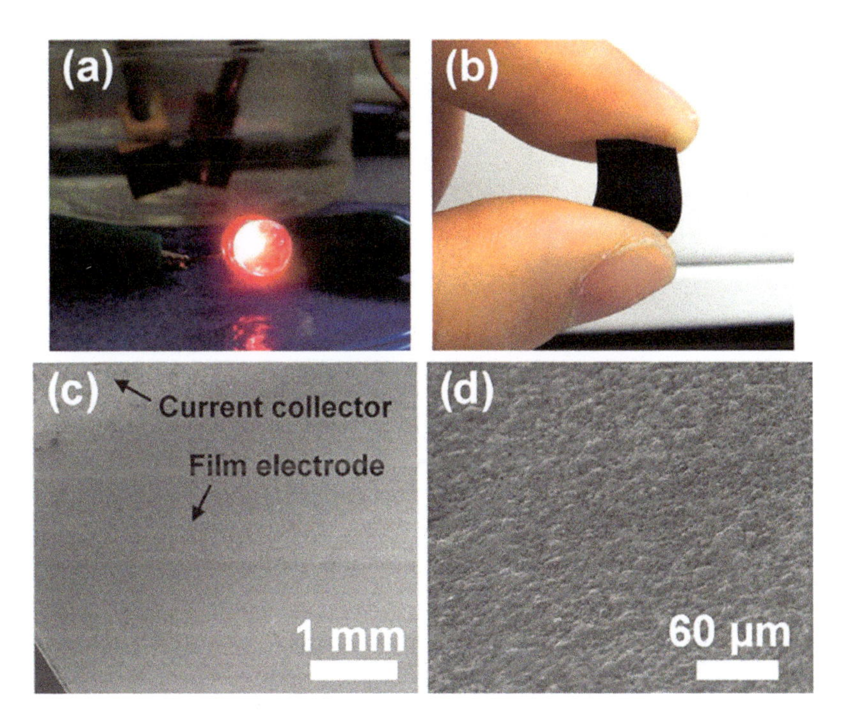

FIGURE 3.8 (a) An optical image showing that a LED was lit brightly even after 300 cycles of the MWNT-α-Fe_2O_3/MWNT (3:1 w/w) supercapacitor, (b) an optical image of a flexible α-Fe_2O_3/MWNT (3:1 w/w) composite anode after 600 cycles, and (c) and (d) SEM images of the top surface of an α-Fe_2O_3/MWNT (3:1 w/w) composite anode after 600 cycles and then bending. The composite anode showed no cracking. Reprinted with permission from the Royal Society of Chemistry [31].

CNTs were accountable for the successful realization of this novel electrode design and good electrochemical performance (50 W h kg^{-1} at 1,000 W kg^{-1}).

Recently, Won et al. synthesized N-rich nanotubes (NRTs) with internal compartments and open mesopores on the surface of nanotubes, and used them in both electrodes in LISCs [32]. A quasi-CNT structure was obtained, with good uniformity in diameter and thin walls (Figure 3.9a and b). Nitrogen doping as well as tin embedment was evidenced by the element map shown in Figure 3.9c. SSA of the prepared NRTs was measured to be approximately 220 m^2 g^{-1}, comparable to CVD-synthesized CNTs. An average pore size of 3.8 nm was identified in their NRTs, which was subject to tuning via the amounts of surfactant templates used during fabrication. These mesopores allow the tin (Sn) source to penetrate through NRT walls during the synthesis of Sn/NRTs anode. They can also function as fast ion-diffusion channels during charge/discharge process. A LISC with NRT cathode and Sn@NRT anode was assembled and tested in a wide potential window of 1.75–4.35 V. Rate capability and capacitance retention over cycling are displayed in Figure 3.9 d and e. The LISC displayed excellent rate performance with near 50% retention of specific capacitance when the current density was increased from 200 mA g^{-1} to 30 A g^{-1}. In terms of energy density based on the mass of both electrodes, a remarkable

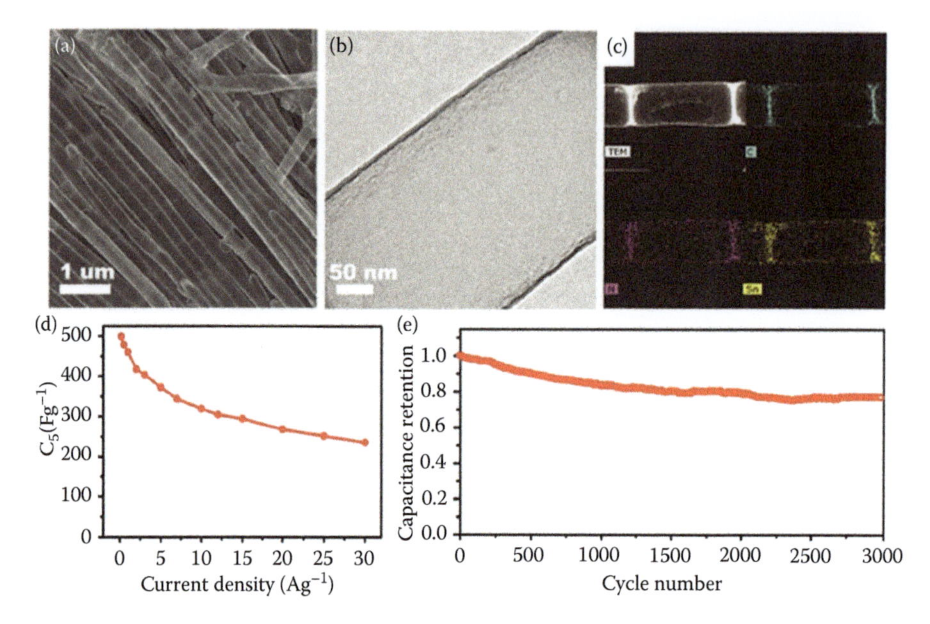

FIGURE 3.9 (a) SEM and (b) TEM images of the NRT, (c) element map image of Sn@NRT structures obtained by TEM, (d) rate capability of the Sn@NRT||NRT hybrid capacitor at different current densities, and (e) cycling performance of the Sn@NRT||NRT hybrid capacitor at a current rate of 2,000 mA g^{-1}. Reprinted with permission from Wiley [32].

energy density of 274 W h kg^{-1} was obtained at 152.5 W kg^{-1}. The LISC maintained 47% of its energy density when the power density was elevated to 22,875 W kg^{-1}. About 80% capacity was retained after 3,000 cycles at a current density of 2 A g^{-1}, showing good long-term capacity retention. N doping not only contributes to the increase of electronic conductivity of nanotube but also enhances electrolyte wettability and creates additional sites for ion adsorption. These advantages, along with the aforementioned unique structure of NRTs, well explain the superior electrochemical performance of the reported LISC.

In addition to graphene and CNT-based cathode materials, some other forms of carbon materials are also studied. Some common strategies that have been embodied in the research efforts include engineering high surface area, introducing heteroatoms, creating 3D or hierarchical structures, partially utilizing graphene or CNTs [2,25,33]. Overall, electric double-layer capacitance plays a major role in carbon-based materials, while pseudocapacitance has been reported in some cases to contribute to the capacitance of a carbon material, e.g., N functional groups [18].

3.5 BATTERY-LIKE CATHODES

Looking beyond the horizon, materials that undergo nonfaradaic process usually can store and release more charged species than EDLC materials that utilize electric double layer formed on the interface of electrode/electrolyte. Transitional metal oxides and conductive polymers have been found to possess excellent pseudocapacitive properties. Mo-based, V-based, Ru-based, etc. oxides have emerged as a new family of electrode materials in SCs. However, most of these materials have relatively low working potentials versus SHE, limiting their application to anode materials in SCs (Figure 3.10) [34]. Li metal oxides, on the other hand, show relatively higher working potentials. Many of these materials are widely used in LIBs as cathode materials, undergoing insertion/desertion reactions. Unlike the carbonaceous nonfaradaic electrodes, insertion-type cathodes have relatively constant working potentials, irrelevant to the state of charge of a cell. Proper mass balancing of insertion-type cathodes and carbonaceous counter electrodes can be an effective alternative to achieving high energy densities, owing to the high operating voltages and high capacities of insertion-type cathodes. These insertion-type cathodes can be categorized based on the crystal structure of each material. Most commonly used are layered oxides. Others include spinel oxides, phosphates, fluorophosphates, and silicates. Several aspects should be taken into consideration in the design of these materials. Compared to EDLC materials, incompetence of operation at high current densities accounts for a major problem for insertion-type cathodes. As the Li insertion reactions not only take place at the surface but also in the bulk of Li metal oxides, polarization may occur when charge/discharge rates increase. Thus, reducing the size and engineering nanostructures can be a viable solution to the diffusion-related issues. In addition, electronic conductivity of Li metal oxides is usually lower than that of carbonaceous materials, which may hinder the rate performance in LISCs. Strategies including surface coating with

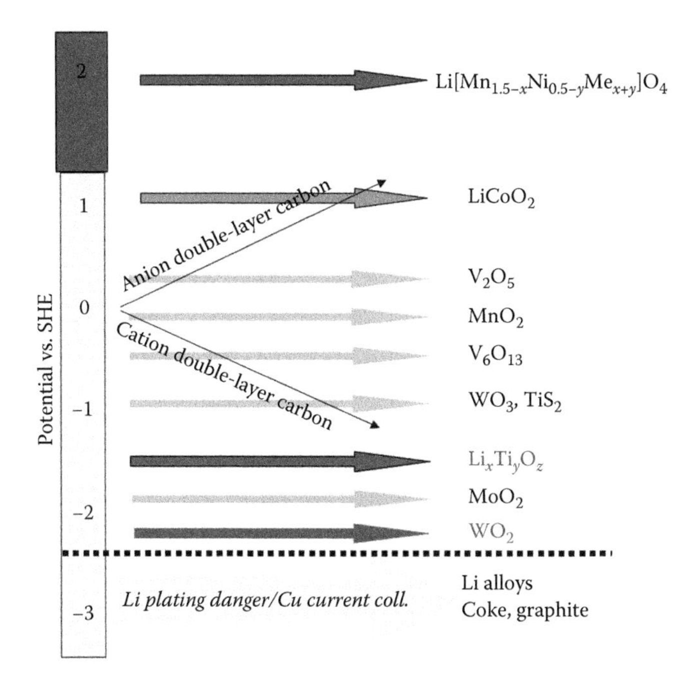

FIGURE 3.10 Plot shows relative potentials of various electrochemical couples versus SHE. Reprinted with permission from Springer [34].

carbon and compositing with conductive materials have proven effective in improving the conductivity at electrode level.

3.5.1 Layered Oxides

Layered oxides have played an important role in the history of LIB development. $LiCoO_2$ was chosen as the cathode material in the first commercial LIBs due to its good stability and high capacity at that time. It was the same group who introduced the LISC concept that built LISCs with $LiCoO_2$ as the cathode in 2002 [35]. However, the structure of this layered oxide collapses when higher voltages are used to extract more Li^+ from it. Mixed metal oxides were proposed to solve this structure problem and deliver higher capacities [36]. A comparative study of $LiNi_{1/3}Co_{1/3}Mn_{1/3}O_2$ was conducted by Yoon et al. in 2008, in which the cathode materials were synthesized via different synthesis methods. AC was used as the anode in the LISCs in this study. $LiNi_{1/3}Co_{1/3}Mn_{1/3}O_2$//AC cells show similar linearity in charge/discharge curves to the cell of AC//AC EDLC, showing capacitive characteristics of the LISC cells (Figure 3.11). The voltage range

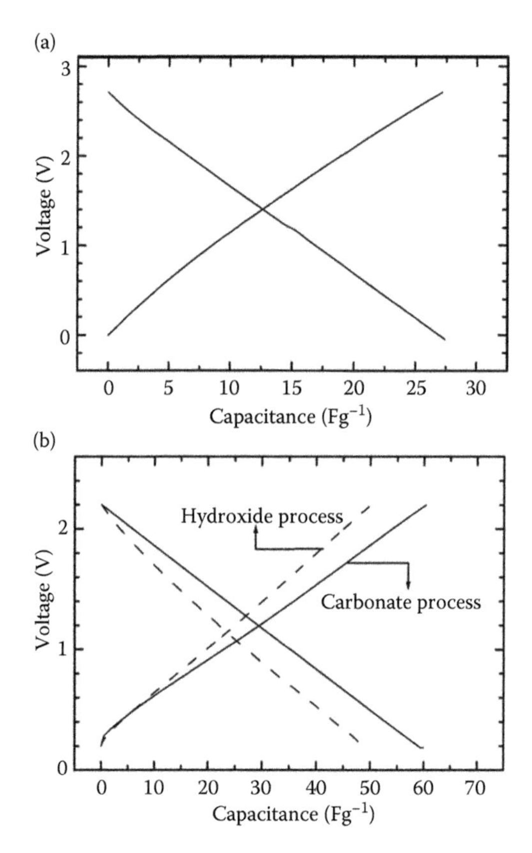

FIGURE 3.11 Charge/discharge curves at a 15 C-rate of (a) AC/AC (EDLC) and (b) $LiNi_{1/3}Co_{1/3}Mn_{1/3}O_2$/AC (asymmetric electrochemical capacitor, AEC) using powders synthesized by both carbonate co-precipitation and hydroxide co-precipitation in the second cycle. Reprinted with permission from Elsevier [41].

was set to be 0.2–2.2 V, relatively lower than those in most AC-cathode LISCs discussed previously. Capacitance of $LiNi_{1/3}Co_{1/3}Mn_{1/3}O_2$//AC cell assembled with cathode via carbonate co-precipitation method achieved 249 F g^{-1} and retained 60 F g^{-1} when high current density of 15 C was used, much higher than the AC//AC EDLC cell. In addition, this LISC retained 94.8% of its capacitance after 500 cycles. As one of the pioneering works adopting Li metal oxides as LISC cathodes, the significance of their results was that they demonstrated the possibility of using battery-like cathodes in LISCs.

Iron was used in $LiNi_{1/3}Mn_{1/3}Fe_{1/3}O_2$ by Karthikeyan et al. to replace the expensive and toxic Co element, and conductive polymers PANi and PPy were separately composited with the layered oxide to enhance its electronic conductivity [37]. Conductive polymers were found to have a significant effect on the capacity of the cathode in a LISC. $LiNi_{1/3}Mn_{1/3}Fe_{1/3}O_2$–PANi/AC cell exhibited a maximum capacity of around 49 W h kg^{-1}, tripled the energy density of LISC with PPy-composited cathode. Only 10% capacity loss was noted after 5,000 cycles for both PANi- and PPy-composited cathodes.

3.5.2 Spinel Oxides

$LiMn_2O_4$ is another widely used LIB cathode material. Mn is more abundant, cheaper, and much less toxic than Co. In addition, the spinel structure of $LiMn_2O_4$ provides 3D Li-ion transport channels for fast Li insertion/desertion. Unlike $LiCoO_2$ that may suffer structural collapse due to Li loss, the structure of $LiMn_2O_4$ is defined by the MnO_4 polyhedral framework and can release all Li during charging. $LiMn_2O_4$, along with $LiCoO_2$ was tested in LISCs by Amatucci group, reported in the same paper [35]. Many efforts have been made investigating $LiMn_2O_4$ cathodes in LISCs. Manganese-based anode, MnO_2/MWCNT, was paired with $LiMn_2O_4$ cathode in a LISC [38]. The LISC delivered a maximum energy density of around 56 W h kg^{-1}, at a power density of 2,400 W kg^{-1}. Hybrid systems of $LiMn_2O_4$/AC and AC/LTO were compared by Cericola et al [39]. When the same power density of 1,200 W kg^{-1} was applied, the $LiMn_2O_4$/AC system delivered an energy density of 45 W h kg^{-1} higher that the AC/LTO system. A hybrid battery-supercapacitor was reported by Hu et al., where the cathode was $LiMn_2O_4$ and AC mixture, and the anode was LTO. In this system, both the cathode and the anode contain battery-like electrode materials [40]. Both capacitive and battery-like charge/discharge behaviors can be observed from the CV curves. Further, advantages from both the battery and SC, i.e., good high capacity of battery and good rate performance of the SC, were embodied by this hybrid device. Over 92% capacity retention was obtained with the configuration of 30% $LiMn_2O_4$–70% AC cathode and LTO anode.

One of the major drawbacks of $LiMn_2O_4$ is that the material suffers from severe capacity decay at elevated temperature due to Mn dissolution. Other transitional metals (Ni, Co, Cr, etc.) have been used to partially substitute Mn to overcome this problem [42]. $LiNi_{0.5}Mn_{1.5}O_4$ is found to be promising and attractive due to the good cycling stability, high working voltage and capacity. There have been intensive research efforts on $LiNi_{0.5}Mn_{1.5}O_4$ in the application of LIBs [43]. Brandt et al. investigated the electrochemcial behavior and degredation mechanism of $LiNi_{0.5}Mn_{1.5}O_4$ in LISCs [44]. It can be oberserved in Figure 3.12a that, in a half

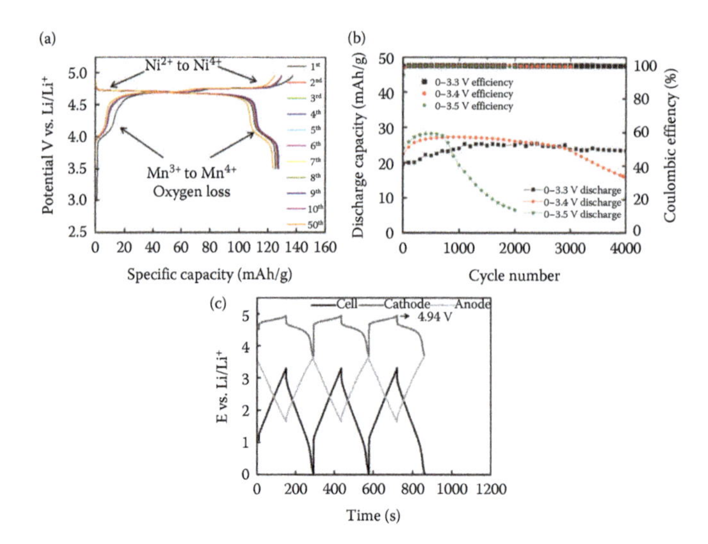

FIGURE 3.12 (a) Charge/discharge profiles of $LiNi_{0.5}Mn_{1.5}O_4$ vs. Li/Li^+ with 1 M $LiPF_6$ in EC:DMC, 1:1 electrolyte in a potential range 3.50–4.95 V vs. Li/Li^+ depicting two potential plateaus above ~4.75 V which correspond to the Ni^{2+} to Ni^{4+} electrochemical reaction via Ni^{3+}. The small electrochemical activity delivered at ~4.0–4.4 V vs. Li/Li^+ is due to the Mn^{3+}/Mn^{4+} electrochemical reaction; (b) variation of discharge capacities and charge/discharge efficiencies with cycle number, at a charge/discharge rate of 10 C for $AC/LiNi_{0.5}Mn_{1.5}O_4$ LISC at 20°C in cell voltage ranges of 0.0–3.3, 0.0–3.4, and 0.0–3.5 V; (c) charge/discharge profiles of the $LiNi_{0.5}Mn_{1.5}O_4/AC$ LISC with a cell voltage of 3.3 V after 3,000 cycles. Reprinted with permission from the Electrochemical Society [44].

cell using Li as the anode, the majority of capacity is occuring at charge/discharge potentials ~4.7 V vs. Li/Li^+, corresponding to conversion between Ni^{2+} and Ni^{4+}, with some contribution from the redox reaction between Mn^{3+} and Mn^{4+}. Full-cell LISCs were assembled with AC as the anode, and voltage windows of 0–3.3, 0–3.4, and 0–3.5 V were applied. It can be concluded that the voltage window has a crucial impact on the long-term cycling stability of the LISCs. Despite that high capacities are shown with LISCs operated at 0–3.4 and 0–3.5 V, LISC operated at 0–3.3 V exhibits negligible capacity decay after 4,000 cycles (Figure 3.12b). Li intercalation/deintercalation of $LiNi_{0.5}Mn_{1.5}O_4$ mostly takes place between 4 and 5 V in a full-cell LISC, contributing to the overall wide voltage window of the LISC (Figure 3.12c). A slight increase of potential plateau is observed after 3,000 cycles, leading to some capacity increase over cycling.

3.5.3 Phosphates

The Goodenough group first examined the electrochemcial behaviors of the olive-phase $LiFePO_4$ [45]. Owing to its excellent thermal stability, high capacity, and environmental friendliness, this material was widely investigated as a promising canditate for LIB cathode [46]. Nevertheless, one-dimensional Li diffusion mechanism in $LiFePO_4$ hinders fast Li transport, presenting an intrinsic drawback of this material. Incorporation of carbon and engineering nanostructured $LiFePO_4$ have been deemed effective approaches tackling the sluggish charge and mass trasnport issue of $LiFePO_4$ [47]. Nano-sized $LiFePO_4$ embedded in a nanoporous carbon matrix was synthesized via sol–gel procedure and subsequent solid state reaction by Wu et al. in 2009 [48]. The composite presented excellent rate performacne in LIB with 60% capacity retention when power density was increased by 50 times from 38 to 1,875 W kg^{-1}. When the material was tested in a LISC with AC as counter electrode, the device showed higher capacity than AC–AC device under the same current density of 6.8 A g^{-1}. The voltage window of their LISC was 0–2.8 V, lower than that of $LiNi_{0.5}Mn_{1.5}O_4$ based system, mainly due to the lower potential plateau of $LiFePO_4$ (3.4 V vs. Li/Li^+). Naoi et al. reported a double-core/shell structured $LiFePO_4$ for the application in LISC [49]. As shown in Figure 3.13, a crystalline $LiFePO_4$ core was covered by an amorphous $LiFePO_4$ layer containing Fe^{3+} defects. Two types of $LiFePO_4$ were denoted as core 1 and core 2, respectively. Another layer of graphitic carbon was used to encapsulate $LiFePO_4$ and form the core-shell structure. The presence of amorphous $LiFePO_4$ greatly enhances the Li-ion diffusivity, while the shell of graphitic carbon significantly improves electronic conductivity of the material, solving two intrinsic

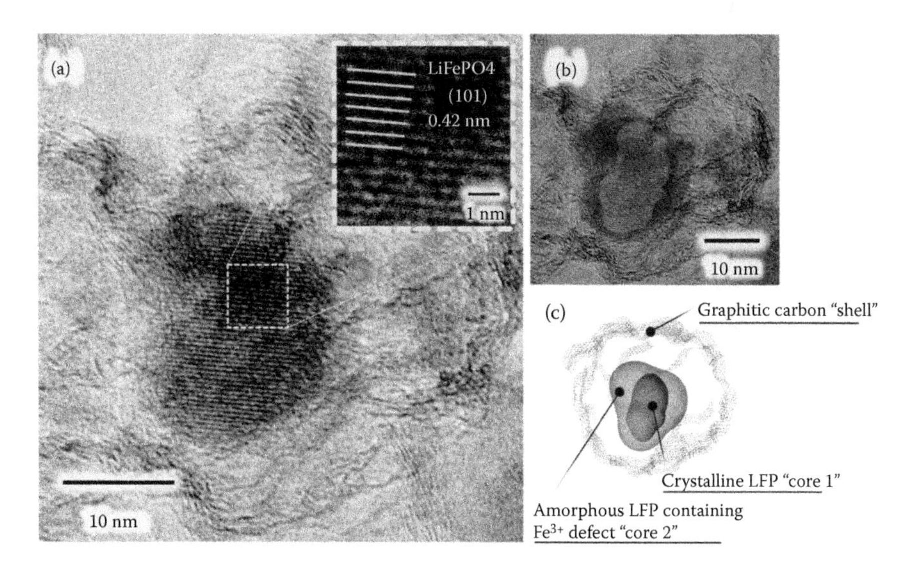

FIGURE 3.13 (a) Higher magnification HRTEM image of the LFP/graphitic carbon composite featuring crystalline LFP nanoparticles (~15 nm) with clear lattice fringes (d = 0.42 nm; LFP(101)) encapsulated within the random graphene fragments derived from high-surface-area KB graphitic carbons; (b) magnified HRTEM image of the LFP/graphitic carbon composite with a marker for each component; (c) schematic illustration of the core–shell nanostructure of the LFP/graphitic carbon composite, representing a minute structure consisting of an amorphous outer sphere of an LFP containing Fe^{3+} defects and an inner sphere of crystalline LFP. Reprinted with permission from the Royal Society of Chemistry [49].

problems with $LiFePO_4$ materials. Current densities of 1–480 C were performed with the $LiFePO_4$ composite in a half cell. Capacities of 60, 36, and 24 mAh g^{-1} were obtained at current densities of 100, 300, and 480 C, respectively. A full-cell LISC was assembled with AC as counter electrode. A voltage range of 0–2.7 V was used, similar to the AC/AC EDLC tested as a comparison group. The full cell demonstrated 30% capacity advantage over the symmetric cell in comparison.

Some other battery-like electrode materials have been exploited in the applciation of LISC cathode, such as fluorophosphates and siliates [50]. It is foreseeable that study of battery-like cathodes in LISCs will continue emerging as more types of LIB cathodes are developed and more techniques are discovered to improve the rate performance of LIB cathodes.

Table 3.2 showcases a number of reported LISCs with different combinations of cathode and anode materials, and electrolyte selections. Cycle performance, energy and power densities, and voltage ranges of these LISCs are summarized.

TABLE 3.2
Typical Examples of Reported LISC Devices and Performance

Device	Electrolyte	Cycle Performance	Energy Density	Power Density	Voltage (V)
CNTs//LTO [52]	Organic $LiPF_6$	92%, 3,000 cycles, 0.65 mA cm^{-2}	4.38 mWh cm^{-3} at 13.5 mW cm^{-3}		0.1–3.0
N-Doped graphene// LTO [53]	Organic $LiPF_6$	64%, 10,000 cycles, 1.5 A g^{-1}	70 W h kg^{-1} at 200 W kg^{-1}; 21 W h kg^{-1} at 8,000 W kg^{-1}		1–3
3DGraphene// Fe_3O_4/G [27]	Organic $LiPF_6$	70%, 1,000 cycles, 2 A g^{-1}	147 W h kg^{-1} at 150 W kg^{-1}; 86 W h kg^{-1} at 2,587 W kg^{-1}		1.0–4.0
AC//$LiNi_{0.5}Mn_{1.5}O_4$ [54]	Organic $LiPF_6$	~81%, 3,000 cycles, 1 A g^{-1}	~19 W h kg^{-1} at ~150 W kg^{-1}; ~8 W h kg^{-1} at ~2.5 kW kg^{-1}		1.5–3.25
$LiNi_{0.5}Mn_{1.5}O_4$//AC [44]	Organic $LiPF_6$	89%, 4,000 cycles, 10 C	~40 W h kg^{-1} at ~1 kW kg^{-1}; 63 W h kg^{-1} at ~100 W kg^{-1}		0–3.3
$Li_3V_2(PO_4)_3$–C//AC [55]	Organic $LiPF_6$	66%, 1,000 cycles	~27 W h kg^{-1} at 255 kW kg^{-1}		0.5–2.75
SGCNT//UC-LTO [56]	Organic $LiBF_4$	–	40–45 W h L^{-1} at 0.1–1 kW L^{-1}; 28 W h L^{-1} at 10 kW L^{-1}		
AC//prelithiated mesocarbon [12]	Organic $LiPF_6$	97%, 1,000 cycles, 2 C	92.3 W h kg^{-1} maximum	5.5 kW kg^{-1} maximum	2.0–4.0
AC//LTO-G [57]	Organic $LiPF_6$	71%, 10,000 cycles, 10 C	50 W h kg^{-1} at 10 W kg^{-1}; 15 W h kg^{-1} at 4,000 W kg^{-1}		~1–2.5
AC//hard carbon Ref [10]	Organic $LiPF_6$	83%, 10,000 cycles, 10 C	60 W h kg^{-1} at ~2,350 W kg^{-1}		1.5–3.9
NAC//Si/C [22]	Organic $LiPF_6$	76.3%, 8,000 cycles, 1.6 A g^{-1}	237 W h kg^{-1} at 867 W kg^{-1}; 141 W h kg^{-1} at 30,127 W kg^{-1}		2.0–4.5
$LiTi_2(PO_4)_3$//AC [58]	Aqueous Li_2SO_4	85%, 1,000 cycles, 10 mA cm^{-2}	24 W h kg^{-1} at 200 W kg^{-1}; 15 W h kg^{-1} at 1.0 kW kg^{-1}		0–1.6

Source: Reprinted with permission from Wiley [51].

3.6 CATHODES OF AQUEOUS LITHIUM-ION SUPERCAPACITORS

Aqueous LISCs have garnered considerable attention due to their special configuration and associated properties. Elimination of organic electrolytes largely alleviates the safety concerns (flammability, toxicity) of LISCs. Cost can be reduced from the replacement of organic solvents with water and less demanding device assembly process. Low-cost lithium salts, such as Li_2SO_4 and $LiNO_3$, can be used as electrolyte salts, rather than the expensive $LiPF_6$ for nonaqueous LIBs and LISCs. However, considering the intrinsic electrochemical stable window of water (usually <1.23 V, H_2 evolution reaction becomes dominant when electrochemical window exceeds 1.23 V). Thus, conventional LIB anodes that have relatively low working potentials vs. Li/Li^+ cannot be utilized in water-based LISCs. Nevertheless, many Li metal oxides and their analogs are not limited by the stability window of water. Conventional layered oxides, spinel oxides, and phosphates are popular candidates for cathode in aqueous LISCs. The Xia group systematically studied hybrid aqueous LISC systems with AC as the anode and Li-intercalated compounds as the cathode [59,60]. Li-doped $Li_{1.1}Mn_2O_4$ was first studied in a neutral Li_2SO_4 aqueous electrolyte and reasonable capacity and cycling performance were obtained. pH and cathode/anode ratio was found to have a significant impact on the utilization of electrode materials and operation voltage window of a LISC. The When the mass ratio of negative/positive electrodes was 2:1 and a neutral electrolyte was utilized, the cell exhibited a maximum operational voltage window of 1.8 V without water splitting (Figure 3.14a). Calculated based on the total mass of active electrode materials, the cell exhibited a specific energy density of 35 W h kg^{-1}, at a power density of 100 W kg^{-1}. EDLC using the same AC for both the anode and cathode only provided 5 W h kg^{-1} at the same power density. Excellent cycling performance was obtained after 20,000 cycles with over 95% capacity retention at a current density of 6 mA cm^{-2}/10 C (Figure 3.14b).

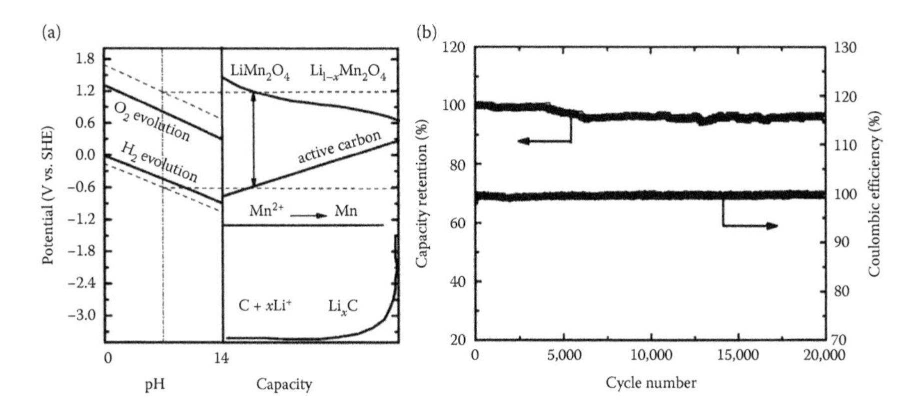

FIGURE 3.14 (a) The potentials of the indicated reactions vs. the standard hydrogen electrode; (b) cycling life of an AC/1 M Li_2SO_4/$LiMn_2O_4$ hybrid aqueous electrochemical supercapacitor at a current rate of 6 mA cm^{-2} between 0 and 1.8 V. Reprinted with permission from the Electrochemical Society [59].

Xia group extended the study to more Li-ion intercalated compounds, comparing $LiMn_2O_4$, $LiCoO_2$, and $LiCo_{1/3}Ni_{1/3}Mn_{1/3}O_2$ in LISCs [61]. Electrolytes with various pH values from 7 to 13 were investigated. Similar maximum specific energy densities were obtained with different LISC cathodes, even though the pH values and the LISC operation windows were different from one to another. $LiMn_2O_4$/AC cell showed the best cycle life and rate performance, while $LiCoO_2$/AC cell steadily decayed over cycling. $LiCo_{1/3}Ni_{1/3}Mn_{1/3}O_2$/AC presented the worst rate performance among the three LISCs. Effect of pH on the capacity fading of $LiCo_{1/3}Ni_{1/3}Mn_{1/3}O_2$, and the possibility of using gel-type water-containing polymer electrolyte was further investigated in LISCs by the same group [62]. Following the pioneering works in aqueous LISCs, many efforts have been devoted into this area. More types of cathode and anode materials have been studied in various aqueous electrolytes [58,63]. However, most of the reported aqueous LISCs have the highest charging voltage below 2 V, due to the intrinsic problems of water-based electrolytes. The voltage limit of aqueous LISCs has greatly deterred the overall energy density, and thus damped the promise of aqueous LISCs.

Recent developments in aqueous LIBs have introduced new opportunities for aqueous LISCs, shedding light on how to widen the voltage window of LISCs. Stable cell operation above 2 V is achieved with "water-in-salt" electrolytes and hydrate-melt electrolytes, both of which contain considerable amount of electrolyte salt in water [64]. High ionic density in electrolytes was found to be responsible for the suppression of water activity and formation of protective SEI [65]. Based on the same idea, a new concept of constructing hybrid SCs was proposed and illustrated in Figure 3.15a. In the proposed SC, traditional aqueous electrolytes are replaced with abovementioned electrolytes: one capacitive electrode and one battery electrode are used, similar to traditional aqueous LISCs. With the widening of operation voltage window, more electrode materials can be utilized, especially the high-voltage and high-capacity cathode materials (Figure 3.15b), providing more options in the design of aqueous LISCs.

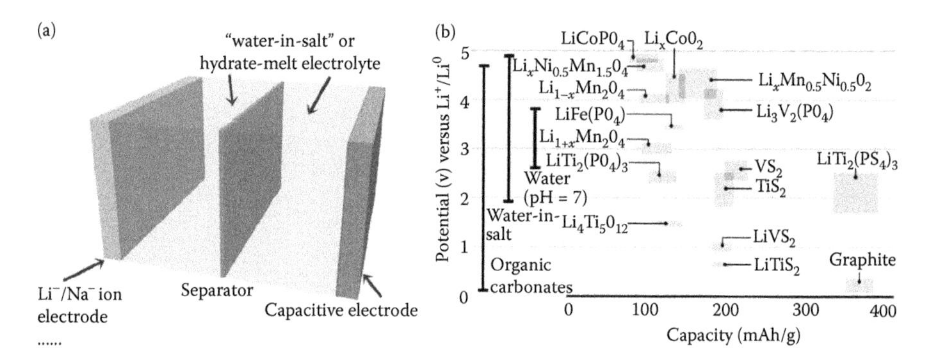

FIGURE 3.15 (a) Conceptual illustration of future Li/Na SCs based on "water-in-salt" or hydrate-melt electrolytes; (b) the Li storage materials system that can be utilized with "water-in-salt" electrolyte. Reprinted with permission from Wiley [51].

3.7 SUMMARY AND OUTLOOK OF LITHIUM-ION SUPERCAPACITOR CATHODES

AC cathode has been employed in a variety of commercial LISCs in market. Yet driven by the intensive demands for high-energy and high-power storage devices, ceaseless research efforts have been made on the improvement of AC cathodes and development of new LISC cathodes. Several strategies have been found to be effective in the design of carbon-based materials, including tuning surface area, controlling pore size and distribution, doping heteroatoms, and creating functional groups. In addition, it is encouraging that new chemistries have come to the center of stage, with emerging efforts on new materials, such as graphene and CNTs. Utilizing a battery-like cathode and a capacitive anode is a viable approach to constructing a LISC. And it is becoming more appealing with the development of new cathode materials that bear high capacity, high voltage, and stable cycling capability. Engineering nanostructures with some battery-type cathodes, e.g., $LiFePO_4$, can significantly enhance the rate capability and ensure the fast charge/discharge capability of the resulted LISCs. These new chemistries have enriched applicable approaches to the achievement of LISC cathodes with higher capacities, shattering the energy-density bottleneck of LISCs due to the relatively low capacity of capacitive cathodes compared to battery-like anodes in traditional LISCs. Further, many of the new cathodes can be applied in aqueous LISCs, which have garnered much interest, especially owing to their safety and cost advantages over organic LISCs.

Exemplified by the research efforts discussed in this chapter, it is justifiable to claim that the energy density gap between batteries and SCs can been greatly narrowed by these LISC systems with new cathodes. Meanwhile, power densities of LISCs can triumph over that of most batteries. These remarkable performances have clearly signaled the possibilities of LISCs being used in a wide range of applications. However, it is important to realize that many of these new materials and methodologies are not readily available yet in wide commercial applications, due to the scalability and cost issues. They can only be commercially acceptable after the costs of these materials can be significantly reduced, which demands further research efforts. Beyond vigorous endeavors on high-energy LISC cathodes, the power requirements for LISCs must be fulfilled for them to be qualified as high-power devices. Moreover, researchers in this field should note that it is misleading to use electrodes with low mass loading of active materials and aim to obtain good performance. Proper and comparable evaluation methods are beneficial to the advancement of this important area of electrochemical energy storage.

REFERENCES

1. C. Sivakumar, J. N. Nian, H. S. Teng, *J. Power Sources* 2005, 144, 295.
2. R. Yi, S. R. Chen, J. X. Song, M. L. Gordin, A. Manivannan, D. H. Wang, *Adv. Funct. Mater.* 2014, 24, 7433.
3. G. G. Amatucci, F. Badway, A. Du Pasquier, T. Zheng, *J. Electrochem. Soc.* 2001, 148, A930; G. G. Amatucci, Badway, F. & DuPasquier, A., in *Intercalation Compounds for Battery Materials* (ECS Proc. Vol. 99) 344–359 (Electrochemical Society, 2000).
4. K. Naoi, S. Ishimoto, Y. Isobe, S. Aoyagi, *J. Power Sources* 2010, 195, 6250.

5. K. Karthikeyan, V. Aravindan, S. B. Lee, I. C. Jang, H. H. Lim, G. J. Park, M. Yoshio, Y. S. Lee, *J. Alloys Compd.* 2010, 504, 224; M. S. Islam, R. Dominko, C. Masquelier, C. Sirisopanaporn, A. R. Armstrong, P. G. Bruce, *J. Mater. Chem.* 2011, 21, 9811.
6. E. Lim, H. Kim, C. Jo, J. Chun, K. Ku, S. Kim, H. I. Lee, I. S. Nam, S. Yoon, K. Kang, J. Lee, *Acs Nano* 2014, 8, 8968; E. Lim, C. Jo, H. Kim, M. H. Kim, Y. Mun, J. Chun, Y. Ye, J. Hwang, K. S. Ha, K. C. Roh, K. Kang, S. Yoon, J. Lee, *Acs Nano* 2015, 9, 7497.
7. T. Brousse, R. Marchand, P. L. Taberna, P. Simon, *J. Power Sources* 2006, 158, 571.
8. P. Simon, Y. Gogotsi, *Nat. Mater.* 2008, 7, 845.
9. V. Khomenko, E. Raymundo-Pinero, F. Beguin, *J. Power Sources* 2008, 177, 643.
10. J. H. Kim, J. S. Kim, Y. G. Lim, J. G. Lee, Y. J. Kim, *J. Power Sources* 2011, 196, 10490.
11. W. J. Cao, J. P. Zheng, *J. Power Sources* 2012, 213, 180; T. Aida, K. Yamada, M. Morita, *Electrochem. Solid-State Lett.* 2006, 9, A534; J. Zhang, X. F. Liu, J. Wang, J. L. Shi, Z. Q. Shi, *Electrochim. Acta* 2016, 187, 134.
12. J. Zhang, Z. Q. Shi, C. Y. Wang, *Electrochim. Acta* 2014, 125, 22.
13. K. Naoi, S. Ishimoto, J. Miyamoto, W. Naoi, *Energy Environ. Sci.* 2012, 5, 9363.
14. A. Jain, V. Aravindan, S. Jayaraman, P. S. Kumar, R. Balasubramanian, S. Ramakrishna, S. Madhavi, M. P. Srinivasan, *Sci. Rep.* 2013, 3, 3002.
15. F. Sun, J. H. Gao, Y. W. Zhu, X. X. Pi, L. J. Wang, X. Liu, Y. K. Qin, *Sci. Rep.* 2017, 7, 40990.
16. B. Li, F. Dai, Q. F. Xiao, L. Yang, J. M. Shen, C. M. Zhang, M. Cai, *Adv. Energy. Mater.* 2016, 6, 18, 1600802.
17. C. H. Kim, J. H. Wee, Y. A. Kim, K. S. Yang, C. M. Yang, *J. Mater. Chem. A* 2016, 4, 4763; H. B. Feng, H. Hu, H. W. Dong, Y. Xiao, Y. J. Cai, B. F. Lei, Y. L. Liu, M. T. Zheng, *J. Power Sources* 2016, 302, 164.
18. C. H. Liu, B. B. Koyyalamudi, L. Li, S. Emani, C. Wang, L. L. Shaw, *Carbon* 2016, 109, 163.
19. S. O. Kim, H. S. Kim, J. K. Lee, *Mater. Chem. Phys.* 2012, 133, 38.
20. S. W. Bokhari, A. H. Siddique, H. Pan, Y. Li, M. Imtiaz, Z. Chen, S. M. Zhu, D. Zhang, *RSC Adv.* 2017, 7, 18926.
21. M. Seredych, D. Hulicova-Jurcakova, G. Q. Lu, T. J. Bandosz, *Carbon* 2008, 46, 1475; E. Raymundo-Pinero, M. Cadek, F. Beguin, *Adv. Funct. Mater.* 2009, 19, 1032; F. Beguin, K. Szostak, G. Lota, E. Frackowiak, *Adv. Mater.* 2005, 17, 2380.
22. B. Li, F. Dai, Q. F. Xiao, L. Yang, J. M. Shen, C. M. Zhang, M. Cai, *Energy Environ. Sci.* 2016, 9, 102.
23. K. Feng, W. Ahn, G. Lui, H. W. Park, A. G. Kashkooli, G. Jiang, X. Wang, X. Xiao, Z. Chen, *Nano Energy* 2016, 19, 187; Y. G. Sun, J. Tang, F. X. Qin, J. S. Yuan, K. Zhang, J. Li, D. M. Zhu, L. C. Qin, *J. Mater. Chem. A* 2017, 5, 13601; Y. F. Ma, H. C. Chang, M. Zhang, Y. S. Chen, *Adv. Mater.* 2015, 27, 5296.
24. H. Zhang, G. P. Cao, Y. S. Yang, *Energy Environ. Sci.* 2009, 2, 932; H. Lee, J. K. Yoo, J. H. Park, J. H. Kim, K. Kang, Y. S. Jung, *Adv. Energy. Mater.* 2012, 2, 976.
25. X. L. Yu, C. Z. Zhan, R. T. Lv, Y. Bai, Y. X. Lin, Z. H. Huang, W. C. Shen, X. P. Qiu, F. Y. Kang, *Nano Energy* 2015, 15, 43.
26. Z. W. Yang, H. J. Guo, X. H. Li, Z. X. Wang, Z. L. Yan, Y. S. Wang, *J. Power Sources* 2016, 329, 339.
27. F. Zhang, T. F. Zhang, X. Yang, L. Zhang, K. Leng, Y. Huang, Y. S. Chen, *Energy Environ. Sci.* 2013, 6, 1623.
28. T. F. Zhang, F. Zhang, L. Zhang, Y. H. Lu, Y. Zhang, X. Yang, Y. F. Ma, Y. Huang, *Carbon* 2015, 92, 106.
29. W. W. Liu, J. D. Li, K. Feng, A. Sy, Y. S. Liu, L. Lim, G. Lui, R. Tjandra, L. Rasenthiram, G. Chiu, A. P. Yu, *ACS Appl. Mater. Interfaces* 2016, 8, 25941.
30. Q. Wang, Z. H. Wen, J. H. Li, *Adv. Funct. Mater.* 2006, 16, 2141.

31. X. Zhao, C. Johnston, P. S. Grant, *J. Mater. Chem.* 2009, 19, 8755.
32. J. H. Won, H. M. Jeong, J. K. Kang, *Adv. Energy. Mater.* 2017, 7, 201601355.
33. R. Gokhale, V. Aravindan, P. Yadav, S. Jain, D. Phase, S. Madhavi, S. Ogale, *Carbon* 2014, 80, 462; H. L. Wang, Z. W. Xu, Z. Li, K. Cui, J. Ding, A. Kohandehghan, X. H. Tan, B. Zahiri, B. C. Olsen, C. M. B. Holt, D. Mitlin, *Nano Lett.* 2014, 14, 1987; M. F. El-Kady, M. Ihns, M. P. Li, J. Y. Hwang, M. F. Mousavi, L. Chaney, A. T. Lech, R. B. Kaner, *Proc. Natl. Acad. Sci. USA* 2015, 112, 4233; R. Wang, J. Lang, P. Zhang, Z. Lin, X. Yan, *Adv. Funct. Mater.* 2015, 25, 2270.
34. I. Plitz, A. DuPasquier, F. Badway, J. Gural, N. Pereira, A. Gmitter, G. G. Amatucci, *Appl. Phys. A: Mater. Sci. Process.* 2006, 82, 615.
35. A. Du Pasquier, I. Plitz, S. Menocal, G. Amatucci, *J. Power Sources* 2003, 115, 171.
36. T. Ohzuku, Y. Makimura, *Chem. Lett.* 2001, 30, 744.
37. K. Karthikeyan, S. Amaresh, V. Aravindan, H. Kim, K. S. Kang, Y. S. Lee, *J. Mater. Chem. A* 2013, 1, 707.
38. S. B. Ma, K. W. Nam, W. S. Yoon, X. Q. Yang, K. Y. Ahn, K. H. Oh, K. B. Kim, *Electrochem. Commun.* 2007, 9, 2807.
39. D. Cericola, P. Novak, A. Wokaun, R. Kotz, *J. Power Sources* 2011, 196, 10305.
40. X. B. Hu, Z. H. Deng, J. S. Suo, Z. L. Pan, *J. Power Sources* 2009, 187, 635.
41. J. H. Yoon, H. J. Bang, J. Prakash, Y. K. Sun, *Mater. Chem. Phys.* 2008, 110, 222.
42. I. Taniguchi, *Mater. Chem. Phys.* 2005, 92, 172; L. Q. Zhang, T. Yabu, I. Taniguchi, *Mater. Res. Bull.* 2009, 44, 707.
43. R. Santhanam, B. Rambabu, *J. Power Sources* 2010, 195, 5442.
44. A. Brandt, A. Balducci, U. Rodehorst, S. Menne, M. Winter, A. Bhaskar, *J. Electrochem. Soc.* 2014, 161, A1139.
45. A. K. Padhi, K. S. Nanjundaswamy, J. B. Goodenough, *J. Electrochem. Soc.* 1997, 144, 1188.
46. M. Konarova, I. Taniguchi, *J. Power Sources* 2010, 195, 3661; Y. G. Wang, P. He, H. S. Zhou, *Energy Environ. Sci.* 2011, 4, 805.
47. J. Nanda, S. K. Martha, W. D. Porter, H. Wang, N. J. Dudney, M. D. Radin, D. J. Siegel, *J. Power Sources* 2014, 251, 8; J. Mun, H. W. Ha, W. Choi, *J. Power Sources* 2014, 251, 386; S. W. Oh, S. T. Myung, S. M. Oh, K. H. Oh, K. Amine, B. Scrosati, Y. K. Sun, *Adv. Mater.* 2010, 22, 4842.
48. X. L. Wu, L. Y. Jiang, F. F. Cao, Y. G. Guo, L. J. Wan, *Adv. Mater.* 2009, 21, 2710.
49. K. Naoi, K. Kisu, E. Iwama, S. Nakashima, Y. Sakai, Y. Orikasa, P. Leone, N. Dupre, T. Brousse, P. Rozier, W. Naoi, P. Simon, *Energy Environ. Sci.* 2016, 9, 2143.
50. V. Aravindan, J. Gnanaraj, Y. S. Lee, S. Madhavi, *Chem. Rev.* 2014, 114, 11619.
51. W. H. Zuo, R. Z. Li, C. Zhou, Y. Y. Li, J. L. Xia, J. P. Liu, *Adv. Sci.* 2017, 4, 7, 1600539.
52. W. H. Zuo, C. Wang, Y. Y. Li, J. P. Liu, *Sci. Rep.* 2015, 5, 7780.
53. Q. Fan, M. Yang, Q. H. Meng, B. Cao, Y. H. Yu, *J. Electrochem. Soc.* 2016, 163, A1736.
54. N. Arun, A. Jain, V. Aravindan, S. Jayaraman, W. C. Ling, M. P. Srinivasan, S. Madhavi, *Nano Energy* 2015, 12, 69.
55. R. Satish, V. Aravindan, W. C. Ling, S. Madhavi, *J. Power Sources* 2015, 281, 310.
56. K. Naoi, W. Naoi, S. Aoyagi, J. Miyamoto, T. Kamino, *Acc. Chem. Res.* 2013, 46, 1075.
57. N. S. Xu, X. Z. Sun, X. Zhang, K. Wang, Y. W. Ma, *RSC Adv.* 2015, 5, 94361.
58. J. Y. Luo, Y. Y. Xia, *J. Power Sources* 2009, 186, 224.
59. Y. G. Wang, Y. Y. Xia, *J. Electrochem. Soc.* 2006, 153, A450.
60. Y. G. Wang, Y. Y. Xia, *Electrochem. Commun.* 2005, 7, 1138.
61. Y. G. Wang, J. Y. Luo, C. X. Wang, Y. Y. Xia, *J. Electrochem. Soc.* 2006, 153, A1425.
62. Y. G. Wang, J. Y. Lou, W. Wu, C. X. Wang, Y. Y. Xia, *J. Electrochem. Soc.* 2007, 154, A228; J. Y. Luo, D. D. Zhou, J. L. Liu, Y. Y. Xia, *J. Electrochem. Soc.* 2008, 155, A789.

63. B. Wang, T. R. Kang, N. Xia, F. Y. Wen, L. M. Chen, *Ionics* 2013, 19, 1527; J. Y. Luo, J. L. Liu, P. He, Y. Y. Xia, *Electrochim. Acta* 2008, 53, 8128.
64. L. M. Suo, O. Borodin, T. Gao, M. Olguin, J. Ho, X. L. Fan, C. Luo, C. S. Wang, K. Xu, *Science* 2015, 350, 938; Y. Yamada, K. Usui, K. Sodeyama, S. Ko, Y. Tateyama, A. Yamada, *Nat. Energy* 2016, 1, 16129.
65. L. Suo, O. Borodin, W. Sun, X. Fan, C. Yang, F. Wang, T. Gao, Z. Ma, M. Schroeder, A. von Cresce, S. M. Russell, M. Armand, A. Angell, K. Xu, C. Wang, *Angew. Chem. Int. Ed.* 2016, 55, 7136.

4 Electrolytes and Separators of Lithium-Ion Supercapacitors

Kunfeng Chen, Wei Pan, Xitong Liang, and Dongfeng Xue
Chinese Academy of Sciences

CONTENTS

4.1 INTRODUCTION

Supercapacitors, including electric double-layer capacitors (EDLCs) and pseudocapacitors, can deliver energy at higher rates than batteries and maintain their specific power for extended periods [1,2]. Inorganic pseudocapacitors consisting of a transition metal oxide/hydroxide electrode and its specific capacitance originate from the redox reaction of active cations with fast charge–discharge rates [3–6]. EDLCs often show lower energy density than that of pseudocapacitors [3,7–9]. However, energy densities of both EDLCs and pseudocapacitors are lower than that of batteries [10,11]. To mitigate the relative disadvantages of EDLCs and pseudocapacitors to realize better performance characteristics, i.e., compatible energy density and power density, employing the concept of integrating both EDLC and Li-ion batteries has emerged and is called lithium-ion supercapacitors (LISCs) [12,13]. The LISC utilizes both faradaic and non-faradaic processes to store charge to achieve higher energy density than EDLC and higher power density than Li-ion batteries without sacrificing cycling stability [14,15]. LISC is expected to bridge the gap between the Li-ion batteries and EDLCs and become the ultimate power source for hybrid electric vehicles (HEV) and electric vehicles (EV) in the near future (Figure 4.1) [16].

While the majority of the attention has been bestowed upon the electrodes, not enough consideration has been presented for the electrolyte and the separator. For LISCs, electrolytes must cater to the needs of both electrodes; hence, in principle, new battery chemistries would have incurred new electrolyte compositions [17]. Electrolyte compositions (especially solvents) are more sensitive to operating potential than capacity of electrodes [18,19]; therefore, as long as the new chemistries operate reasonably within the electrochemical stability window of the electrolytes, major changes in the skeleton composition are not mandatory [17,20].

The ideal separator has the characteristics of high ionic conductivity, high electrical resistance, and satisfactory stability in electrolyte solutions [21]. Currently, the most widely utilized membrane is the commercialized microporous polyolefin films. The closely ordered micropores of the separator, which act as a passage for electrolyte flooding through the membrane, give it a high ionic conductivity [22,23]. However, polypropylene separator penetrates most of flowing materials in the electrolyte through pores between layered-fibrous structures. In this chapter, we provided insights into the roles and progresses of electrolyte and separator in the LISC. With the objective of improving the electrochemical performance, we dedicated strategies to stabilize electrolyte and

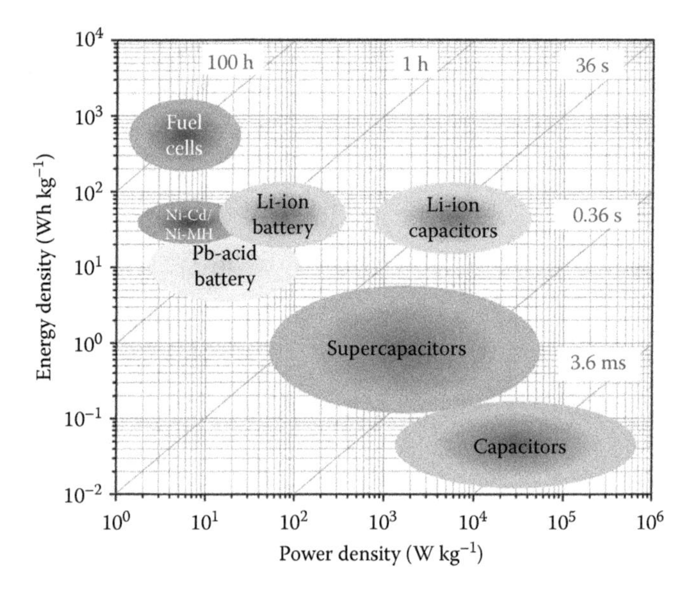

FIGURE 4.1 Ragone plot shows the demand for high power/energy electrochemical energy storage devices relative to present day technology [12]. Reprinted with permission from Ref. [12]. Copyright 2014, American Chemical Society.

separator during charge–discharge cycling. Using novel electrolytes allows the increase of the cell voltage. Full understanding of their electrochemical mechanisms can be achieved only by utilizing advanced characterization tools over multiple length scales from electrode materials to electrolyte and separator as well as their interaction.

4.2 ELECTROLYTE TYPES

The electrolytes of LISCs usually consist of solvents, lithium salts, and additives [24,25]. The role of electrolytes is to serve as the medium for the transfer of charge, which are in the form of ions, between anode and cathode, which are in close interaction with both electrodes [26,27]. The electrolytes often include aqueous and nonaqueous electrolytes. The interfaces between the electrolyte and electrodes often dictate the performance of devices. A generalized list of the minimal requirements of electrolytes should include the following [24]: (1) a good ionic conductor and electronic insulator. Ion transport can be facile and self-discharge can be kept to a minimum; (2) a wide electrochemical window. An electrolyte degradation would not occur within the range of the working potentials of both the cathode and the anode; (3) inertness to other cell components such as cell separators, electrode substrates, and cell packaging materials; (4) good resistance against various abuses, such as electrical, mechanical, or thermal ones; and (5) environmentally friendly nature.

4.2.1 Aqueous Electrolytes

Compared with organic electrolytes, aqueous electrolytes have smaller solvated ions, higher ionic concentration, and higher ionic conductivity, which own some additional advantages such as low cost, nonflammability, low viscosity, and excellent safety [28,29]. Therefore, supercapacitors commonly provide much higher specific capacitances and power densities in aqueous electrolytes than those in organic electrolytes. Moreover, aqueous electrolytes can be easily produced on a large scale and used without the need of particular conditions, while hygroscopic organic ones usually need strict conditions and complicated processes to avoid the introduction of moisture [30]. Generally, the selection criteria for electrolytes are based on the sizes and radii of both hydrated cations and anions as well as the ionic conductivity (Table 4.1) [31]. The most commonly used aqueous electrolytes are KOH, H_2SO_4, and Na_2SO_4(Li_2SO_4) for alkaline, acidic, and neutral solutions, respectively [32,33]. However, the disadvantage of aqueous electrolytes is their narrow voltage window (~1.0 V) due to the thermodynamic decomposition of water at 1.23 V and low overpotential for H_2 and O_2 evolution [34,35]. In order to effectively address this problem, considerable effort has been made to improve the maximum voltage window of aqueous solution, for e.g., by mixing cation electrolyte [36,37].

In addition, the internal resistance of the cell can also be decreased in concentrated electrolytes, resulting in large power delivery. However, the highly corrosive electrolytes (strong acid and basic) will increase the fabrication costs and device weights owing to using corrosion-resistant current collectors (such as Au, Pt, Ti, graphite sheets, and conductive carbon paper) [32]. Moreover, the concentrate electrolytes will give rise to high self-discharge rate [38,39]. Aqueous electrolytes should be deaerated

TABLE 4.1
Radii of Hydrated Ions and Ionic Conductivity Values

Ions	Hydrated Ion Radii (nm)	Ionic Conductivity (S cm^2 mol^{-1})
H^+	0.282	349.65
Li^+	0.382	38.66
Na^+	0.358	50.08
K^+	0.331	73.48
NH_4^+	0.331	73.50
OH^-	0.300	198.00
Cl^-	0.332	76.31
ClO_4^-	0.338	67.30
NO_3^-	0.335	71.42
SO_4^{2-}	0.379	160.00

before the electrochemical measurements to get rid of the dissolved oxygen. Recently, redox-active electrolytes have attracted more attention due to the active components, which can contribute additional pseudocapacitance to the overall capacitance of supercapacitors [40–42]. The commonly used electrochemically redox-active compounds are hydroquinone, *m*-phenylenediamine, KI, lignosulfonates, etc.

4.2.2 Nonaqueous Electrolytes

Due to the low voltage window of aqueous solutions, the requirements of high voltage and energy density have turned our attention to organic electrolytes. Compared with aqueous electrolytes, organic electrolytes can provide a voltage window as high as 2.5–3.0 V, giving rise to energy density 6–9 times that of the aqueous electrolytes [43]. Generally, the operating voltage is set to 2.5 V to inhibit the electrolyte from oxidation due to over-charging. The possibility to increase the operating voltage of EDLCs containing acetonitrile (AN) and propylene carbonate (PC)-based electrolytes has been intensively investigated in the past years [44–47]. These studies showed fundamental limitations to apply voltage beyond 3 V while maintaining stable performance when ANs are used in combination with the state-of-the-art electrolytes. In the case of AN, the electrolyte degradation seems to be the main obstacle for the increment of operating voltage, and the positive electrode seems to be the one that suffers the most in the case of such an increase [48]. In the case of PC, the formation of carbonates and the evolution of gaseous species (H_2 and CO_2) have been identified as the main reasons for the decrease of performance at high voltages [45]. Considering the limitations, and with the aim to increase the operating voltage of EDLCs, several electrolytes have been proposed in the past as alternatives to the AN- and PC-based electrolytes. Generally, the alternative electrolytes proposed so far can be divided in two main categories: ionic liquids (ILs) and organic solvent-based electrolytes (Table 4.2).

The commonly used ILs for the applications in supercapacitor are imidazolium, pyrrolidinium as well as asymmetric, aliphatic quaternary ammonium salts with anions such as tetrafluoroborate, trifluoromethanesulfonate, bis(trifluoromethanesulfonyl)

TABLE 4.2
Comparison of the Operating Voltage of Organic and Ionic Liquids Based Electrolytes

Organic Electrolytes		Operating Voltage (V)	Conductivity (mS/cm)
Conventional	Acetonitrile (AN)	2.7	55.0
	Propylene carbonate (PC)	2.8	13.0
Alternative	Linear sulfones	3.2	n.a.
	Alkylated cyclic carbonate	3.2	n.a.
	Adiponitrile (ADN)	3.6	4.3
Li-ion battery	EC-DMC 1M $LiPF_6$	3.0	11.8
	Ionic liquids (IL)		
Solvent-free	Imidazolium-based	3.2	10.0
	Pyrrolidinium-based	3.7	2.6–3.8
	IL-mixtures	3.0	4.9
Co-sol.	IL-PC mixtures	3.0	n.a.

EC, ethylene carbonate; DMC, dimethyl carbonate; $LiPF_6$, lithium hexafluorophosphate.

imide, bis(fluorosulfonyl)imide, or hexafluorophosphate [49]. However, ILs have some disadvantages such as relatively high cost, high viscosities, and low ionic conductivity at room temperature [50,51]. The ionic conductivity of ILs at room temperature is just several mS cm^{-1}; thus, they are commonly employed at higher temperatures, still failing to meet the requirements for supercapacitors in the temperature ranges of –30°C to 60°C in practical applications. Similar to organic electrolytes, ILs are also hygroscopic and must be handled in an inert atmosphere. Therefore, the design of ILs with a wide potential range, high conductivity in a wide temperature range, and low cost is still challenging, and will be of great significance in applying high-performance, safe, and green energy storage devices [31].

4.3 SEPARATOR TYPES AND THEIR PHYSIOCHEMICAL PROPERTIES

Separators with a rich pore structure are one of the key components of LISCs and other energy storage devices. The function of separators is to block the contact of positive electrode and negative electrode and transfer ions only [52,53]. To reach the demands of high power density for LISCs, separators must be used at the right environment that match with their physiochemical properties. Separators play two main roles during the cell operation: (1) resting between the anode and the cathode to prevent internal short circuiting, and (2) providing a path for ionic conduction in the liquid electrolyte throughout the interconnected porous structure [54]. Separators for the liquid electrolyte are currently engineered as porous membranes, nonwoven mats, or multilayers consisting of porous membranes and/or nonwoven mats. Table 4.3 lists the major separator manufacturers, showing that most of the commercial separators are made of porous polyolefin membranes [55].

TABLE 4.3
Properties of the Commercial Separators

Separator	Separator Materials	Thickness (μm)	Gurley Number (s)
Celgard®2325 (Polypore)	PP/PE/PP	27	570.0
Celgard®2500 (Polypore)	PP	27	180.1
FS2190 (Freudenberg and Co. KG)	PP	176	—
Polyamide 0.2 μm (Sartotius Stedium Biotech GmbH)	Polyamide	116	31.9
Copa Spacer (Spez. Papierfabrik Oberschmitt GmbH)	Cellulose	50	6.5
Separion® (Evonik)	Ceramic on PET	28	22.8
GF/C (Whatman)	Glass fiber	283, 359	1.0, 2.3

Source: Reprinted with permission from Ref. [55]. Copyright 2014, MDPI.

An ideal separator should have an infinite electronic but a zero ionic resistance. In practice, the electrical resistivity of the polymers used for separators is on the order of 10^{12}–10^{14} Ω/cm, i.e., they are electrical insulators [56–58]. In the meantime, a low internal ionic resistance is especially important for HEV/EV applications where a LISC needs to be able to offer high power. However, the existence of the separator always increases the ionic resistance of the inter-electrode medium (consisting of the separator and the liquid electrolyte) because (1) the finite porosity of a separator implies a restricted contact area between the electrolyte and the electrodes; and (2) the tortuosity of the open porous structure results in a longer mean path for the ionic current compared to when the liquid electrolyte is used alone [59,60]. Generally, a thin membrane with a high porosity and a large mean pore size can minimize the ionic resistance, enabling high-specific power. However, too great of a porosity and a small membrane thickness can reduce the mechanical strength of the membrane and increase the risk of inner battery electrical shorting. In practice, most separators for thr liquid electrolyte in use today are 20–30 μm thick, have submicron-sized pores, and possess porosity ranging from 40% to 70% [61,62]. In addition, separators should be mechanically strong, with no skew or yield, to keep the anode and cathode from contacting each other during the whole device lifetime. Separators also must possess dimensional stability at elevated temperatures, especially for applications in high-power devices. The specifications for a Celgard separator are listed in Table 4.4.

Separators in the liquid electrolyte can be divided into six types: microporous films, nonwovens, ion exchange membranes, supported liquid membranes, solid polymer electrolytes, and solid ion conductors [63,64]. Recently, polymer electrolytes and solid ion conductors have been developed served as both the separator and the electrolyte. Polymer electrolytes (e.g., poly(ethylene oxide), poly(propylene oxide)) have attracted considerable attention for batteries in recent years [65]. These polymers form complexes with a variety of alkali metal salts to produce ionic conductors that serve as solid electrolytes. Their use in batteries is still limited due to poor electrode/electrolyte

TABLE 4.4
Celgard Requirements for Li-Based Separators

Membrane name	Celgard 2325
Process	Dry
Composition	PP–PE–PP
Thickness (μm)	25
Porosity (%)	41
Pore size (μm)	0.09×0.04
Gurley value (s) per 100 cm^3	575
Tensile strength (MD) (kg cm^{-2})	1900
Tensile strength (TD) (kg cm^{-2})	135
Melting temperature (°C)	134/166
Thermal shrinkage (%)	2.5

Source: Reprinted with permission from Ref. [62]. Copyright 2007, Elsevier.

interface and poor room temperature ionic conductivity. Solid ion conductors (ceramic membrane) allow one or more kinds of ions to migrate through their lattice when a potential gradient or a chemical gradient is present [66]. Rigid ceramic membranes are gaining more and more market share due to their special properties such as high chemical/electrochemical stability and thermal load. However, the complex fabrication process of these materials makes the final products expensive.

4.4 CORRELATION BETWEEN ELECTROLYTE PROPERTIES AND LITHIUM-ION SUPERCAPACITOR PERFORMANCE

The aqueous electrolytes show higher ionic concentration and higher ionic conductivity, which can satisfy the high power of LISC [67]. The electrochemical performances of $LiMn_2O_4$ which served as electrode materials of LISC in $LiNO_3$ electrolyte and organic electrolyte are shown in Figure 4.2 [68]. The results showed that $LiMn_2O_4$ supercapacitors in $LiNO_3$ electrolyte had similar discharge capacity and potential window (1.2 V) as that of organic electrolytes (Figure 4.2). In $LiNO_3$ aqueous electrolyte, the reaction kinetics of $LiMn_2O_4$ supercapacitors was very fast. Even, at current densities of 1 and 5 A/g, aqueous electrolyte gave good capacity compared with organic electrolyte at a current density of 0.05 A/g.

Besides the improvement of electrode materials, another alternative strategy to largely improve the specific capacitance was the introduction of redox materials into electrolyte [69]. These redox materials can be mainly divided into two categories: organic ring compounds with unsaturated bonds and inorganic ionic compounds with variable valence metal ions [70–72]. A gel polymer electrolyte (PVA–H_2SO_4–P–benzenediol) and activated carbon electrodes have been fabricated a supercapacitor [73]. Due to the faradic pseudocapacitance induced by *p*-benzenediol/*p*-benzoquinone in the electrolyte, the supercapacitor exhibits a large specific capacitance of 474 F g^{-1} and a high energy density of 11.31 Wh kg^{-1}. With the *p*-phenylenediamine (PPD)

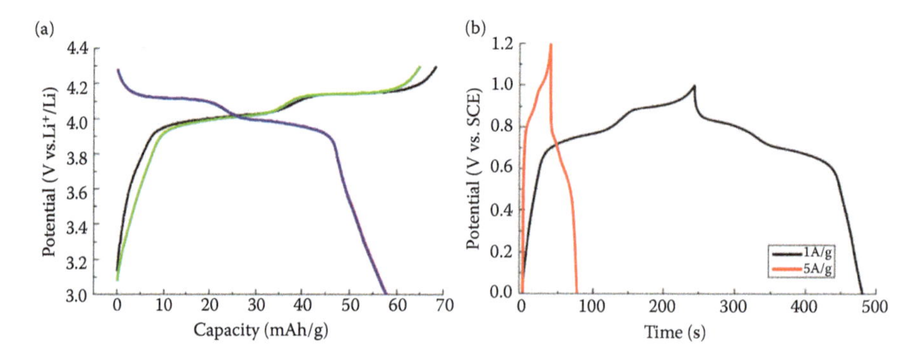

FIGURE 4.2 CV curves and galvanostatic charge–discharge curves of $LiMn_2O_4$ in (a) organic electrolyte (1 M $LiPF_6$ in EC/DMC/DEC, 1:1:1vol%) and (b) 1 M $LiNO_3$ electrolyte [68]. Reprinted with permission from Ref. [68]. Copyright 2014, Elsevier.

addition into lithium perchlorate organic electrolyte, the cell capacitances was 69 F g^{-1}, largely improved in comparison to 25 F g^{-1} without PPD (Figure 4.3) [70].

By combining electric double-layer capacitance and pseudocapacitance of redox-electrolyte $K_3Fe(CN)_6$, the specific capacitance could be increased five fold compared with the conventional electrode–electrolyte system (Figure 4.4) [74]. This large improvement is attributed to the additional redox reactions on the graphene paper electrode via the constituent ions of $K_3Fe(CN)_6$ in the redox-electrolyte. More importantly, the potential interval could reach as high as 1.6 V beyond the limited operating voltage of water (~1.23 V). After 5000 continuous cycles, 94% of the initial capacitance was retained. This designed graphene-paper-electrode/redox-electrolyte system provides a versatile strategy for high capacitance supercapacitor systems.

Furthermore, our group designed a new type of ionic pseudocapacitor system with excellent contributions from ionic-state redox mediators, including a redox couple $[Fe(CN)_6]^{3-}/[Fe(CN)_6]^{4-}$ in the electrolyte and redox cations in highly electroactive colloid electrodes [75]. The highest pseudocapacitance value of 12,658 mF cm^{-2}

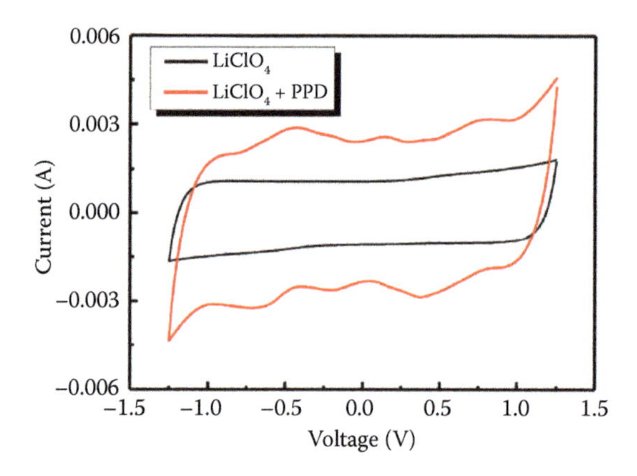

FIGURE 4.3 The comparison of CV curves at 10 mV s^{-1} in $LiClO_4$ and $LiClO_4$ + PPD electrolytes [70]. Reprinted with permission from Ref. [70]. Copyright 2014, Elsevier.

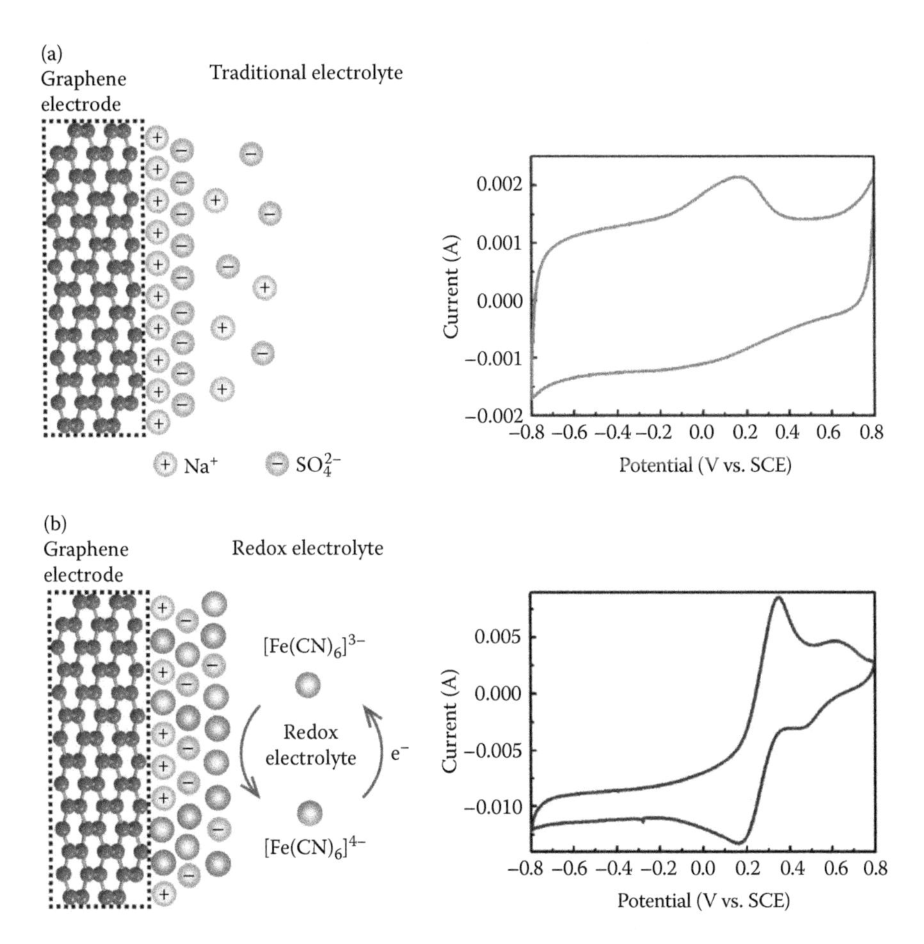

FIGURE 4.4 Charge storage mechanisms of the graphene paper electrode in the traditional electrolyte system (a) and the redox-electrolyte system (b). (a) Graphene paper electrode in the neutral Na_2SO_4 electrolyte, which displays electric double-layer capacitance with quasi-rectangular CV curves. (b) After adding redox-active $K_3Fe(CN)_6$ to the Na_2SO_4 electrolyte, this supercapacitor system combines two charge storage mechanisms, pseudocapacitance and EDLC. The novel electrode and electrolyte system can significantly increase the specific capacitance of supercapacitors [74]. Reprinted with permission from Ref. [74]. Copyright 2015, Royal Society of Chemistry.

was obtained, which is a sevenfold increase in the specific capacitance of the $CoCl_2$ electrode for $K_3Fe(CN)_6$ in a KOH alkaline electrolyte at a current density of 20 mAcm^{-2} and at a potential interval of 0.55 V. The currently designed supercapacitor systems show a versatile strategy to design high performance supercapacitors with $CoCl_2$, $CuCl_2$, $NiCl_2$, and $FeCl_3$ electrodes. In the ionic pseudocapacitor system, the cations were transformed to electroactive hydroxide colloids by electric field-assisted chemical precipitation in a KOH aqueous solution [76–78]. The presence of $[Fe(CN)_6]^{3-}/[Fe(CN)_6]^{4-}$ couple in the redox electrolyte can be proven by FTIR (Figure 4.5).

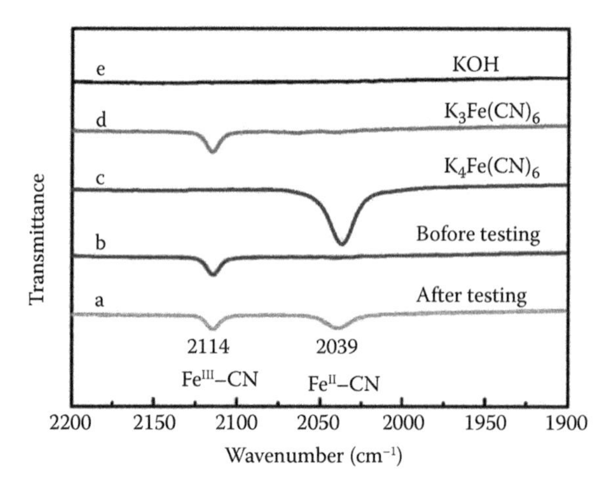

FIGURE 4.5 Fourier transform infrared spectra (FTIR) of $K_3Fe(CN)_6$ and KOH aqueous electrolyte. Infrared spectra of the mixing electrolyte including 0.3 M $K_3Fe(CN)_6$ and 2 M KOH after (a) and before (b) the electrochemical test, and (c) 0.3 M $K_4Fe(CN)_6$, (d) 0.3 M $K_3Fe(CN)_6$, and (e) 2 M KOH aqueous solution. The infrared stretching frequency of Fe^{III}–CN was found to be 2114 cm^{-1}, while the stretching frequency of Fe^{II}–CN was 2039 cm^{-1} [75]. Reprinted with permission from Ref. [75]. Copyright 2014, Royal Society of Chemistry.

Specifically, electroactive materials in the electrode can be reacted with electrolyte to form high-performance supercapacitors [79–81]. High electroactive Ni-based colloidal electrode materials have been synthesized by in situ electrochemical activation of $NiCl_2$ electrode in the alkaline electrolyte (Figure 4.6) [79]. The highest specific capacitance of the activated Ni-based electrodes is 10,286 F/g at the current density of 3 A/g, indicating the three-electron Faradic redox reaction ($Ni^{3+} \leftrightarrow Ni$) occurred in these activated electrodes. Figure 4.7a shows the evolution of open-circuit potential (OCP) of $NiCl_2$ electrode when putting it in KOH electrolyte. OCP increases with the prolonging of time to 2 h, indicating the occurrence of ion exchange reactions to form $Ni(OH)_2$. OH^- ions were exchanged with Cl^- ions to form $Ni(OH)_2$ within the initial $NiCl_2$ electrode.

$$NiCl_2 + OH^- \rightarrow Ni(OH)Cl + Cl^- \tag{4.1}$$

$$Ni(OH)Cl + OH^- \rightarrow Ni(OH)_2 + Cl^- \tag{4.2}$$

Electrochemical impedance spectra (EIS) also prove the formation of different materials during this process (Figure 4.7b). When we conducted cyclic voltammetry (CV) cycling to the as-obtained $NiCl_2$ electrode, the electrical-field induced reactions can occur (Figure 4.7c).

$$Ni(OH)_2 + OH^- \rightarrow Ni_x{}^{II}Ni_{(1-x)}{}^{III}OOH + H_2O + e^- \tag{4.3}$$

$$Ni_x{}^{II}Ni_{(1-x)}{}^{III}OOH \rightarrow Ni^{III}OOH + e^- \tag{4.4}$$

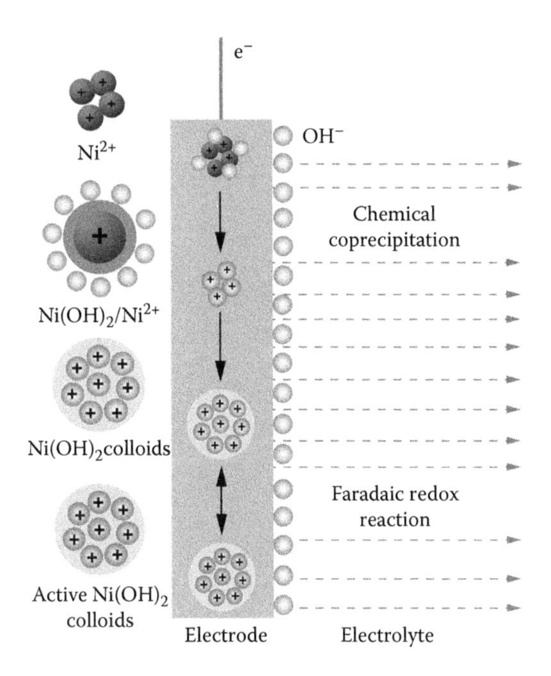

FIGURE 4.6 Schematic showing the reaction mechanism of a Ni-based pseudocapacitor system. Black arrows show the detailed information of phase transformation of metal cations within electrode. Red arrows represent electric field [79]. Reprinted with permission from Ref. [79]. Copyright 2016, Royal Society of Chemistry.

$$Ni^{III}OOH + H_2O + e^- \rightarrow \text{activated-}Ni(OH)_2 + OH^- \quad (4.5)$$

First, $Ni(OH)_2$ materials were oxidized into the high oxidation state of $Ni_x{}^{II}Ni_{(1-x)}{}^{III}OOH$ or $Ni^{III}OOH$ products, and then the charged products were reduced into high activated α-$Ni(OH)_2$ colloidal materials. After five CV cycles, the activated process of Ni-based electrode materials was stopped; thus, high-activity pseudocapacitive electrode materials were formed. Different activation methods have proven the importance of electrochemical activation. The capacitance characteristics of the Ni-based electrode electrodes mainly originate from reversible faradaic redox reactions as follows [82,83]:

$$Ni(OH)_2 + OH^- \leftrightarrow NiOOH + H_2O + e^- \quad (4.6)$$

According to this redox mechanism, the specific capacitance is limited because only one electron was transferred. However, the three-electron redox reaction can occur in the designed Ni-based colloidal supercapacitor system:

$$Ni^{3+} + e^- \leftrightarrow Ni^{2+} + 2e^- \leftrightarrow Ni \quad (4.7)$$

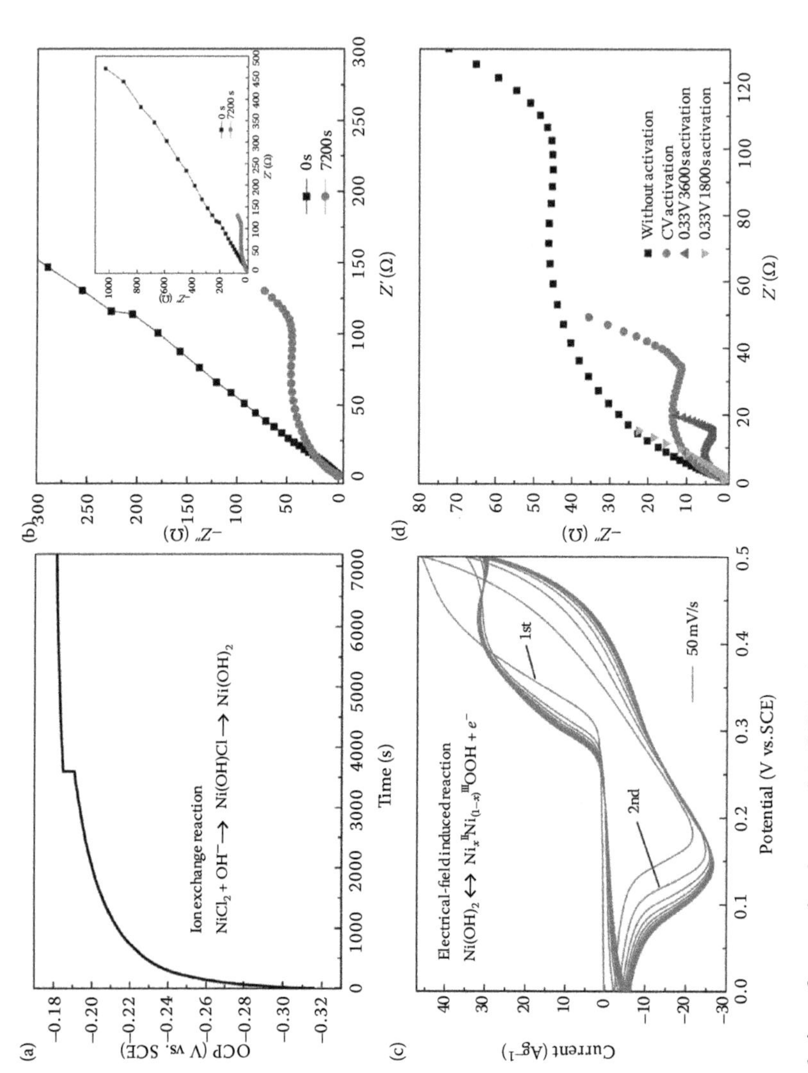

FIGURE 4.7 (a) Evolution of open-circuit potential (OCP) with time in KOH electrolyte. (b) EIS spectra of $NiCl_2$ electrode at 0 and 7200 s at OCP state. (c) CV activation process from the initial $NiCl_2$ in KOH solution. (d) EIS spectra of $NiCl_2$ electrode after undergoing CV activation and constant potential activation [79]. Reprinted with permission from Ref. [79]. Copyright 2016, Royal Society of Chemistry.

Upon potential cycling and constant potential activation, the decrease of charge transfer resistance can be found. The activation and utilization of multiple-electron reactions is an efficient route to increase the energy density of supercapacitors. This newly designed colloidal pseudocapacitor is compatible with inorganic pseudocapacitor chemistry, which enables us to directly use metal cations via commercial salts instead of their condensed matters such as oxide/hydroxide materials.

In the organic electrolyte, electrode materials can also be reacted with organic electrolyte to form specific structure, i.e., SEI film, which determines the electrochemical performance of LISC [84,85]. The existence of a Cu^+-ion long-range transfer path was identified at the potential widow of 1.30–1.60 V during both charging and discharging processes [86]. TEM images show that these nanowire networks hanging CuO nanoparticles provide a Cu^+ diffusion path within the designed CuO/C integrated anode (Figure 4.8). These nanowire networks served as Cu^+ long-range diffusion paths, because only the Cu_2O phase can be formed at this stage. This provides new insights into the conversion reaction of inorganic anode materials and

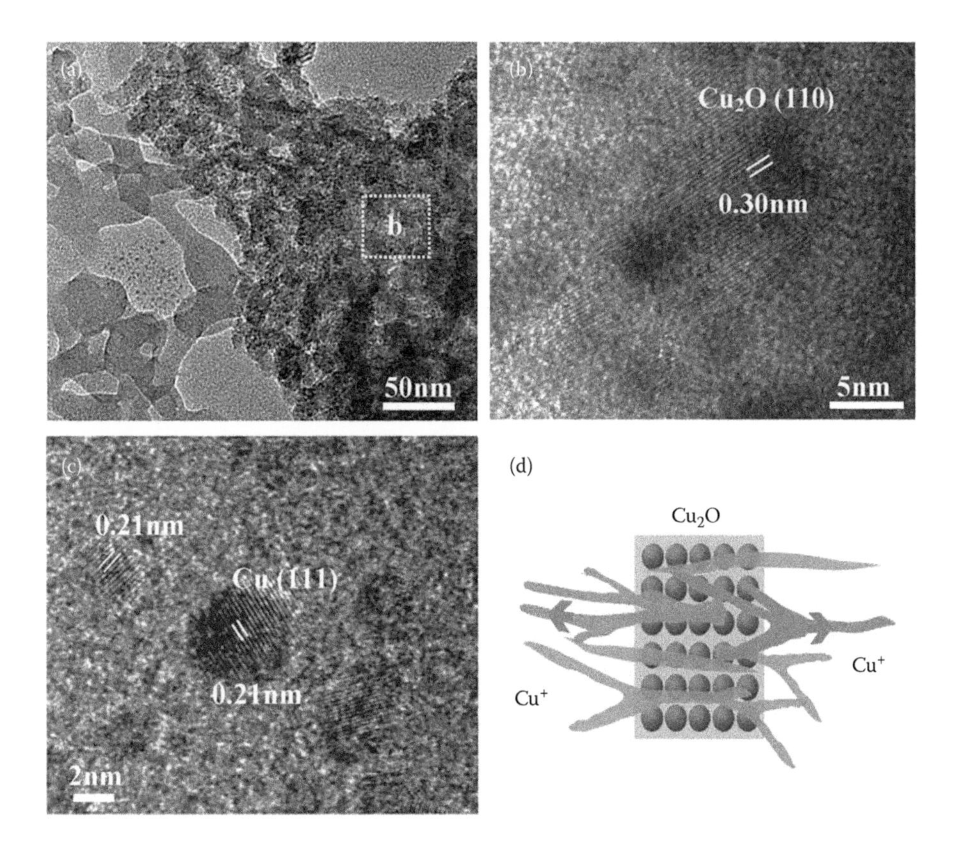

FIGURE 4.8 (a–c) TEM and HRTEM images of CuO/Cu integrated electrode after discharged to 1.6 V. (d) Schematic illustration shows the formation of nanowire networks among CuO microsheets, which serve as Cu^+ ion long-range transfer path [86]. Reprinted with permission from Ref. [86]. Copyright 2014, Royal Society of Chemistry.

can favor the development of high-performance conversion anodes for lithium-ion batteries. The parasitic reactions of electrodes with liquid electrolytes are complex, which needs rationally designing of electrolytes and electrode/electrolyte interfaces for stable operation of LISCs.

4.5 CORRELATION BETWEEN SEPARATOR PROPERTIES AND LITHIUM-ION SUPERCAPACITOR PERFORMANCE

The current polyolefin commercial separator cannot meet the increasing requirements of high power density, high energy density, and high safety performance of LISCs, due to its inherent shortages, i.e., serious thermal shrinkage and low puncture strength [87–89]. Degussa, the German company, first proposed the concept of ceramic separator, i.e., the use of organic material flexibility and good thermal stability of inorganic materials to improve the overall performance of the diaphragm [90]. The ceramic separator is shown in Figure 4.9. When the organic membrane completely melted, inorganic coating still maintained the integrity of the separator to prevent the occurrence of positive and negative electrode short-circuit phenomenon. The separator has the advantages of good wettability, permeability, and mechanical strength, as well as high-temperature thermal stability, and the melting temperature can reach 210°C. However, the limitation of this technique is that the dehydration condensation reaction can occur in the range of 180°C–200°C, which is the temperature of the formation of the silicon–oxygen coupling agent. Therefore, in order to use the technology, the heat-resistant temperature of the separator-substrate material should be higher than 200°C.

Recently, the inorganic particles coating has been developed to improve the thermal stability of commercial polyolefin separators [91–93]. The most common inorganic particles, such as SiO_2, Al_2O_3, have been used to coat polyolefin separator [94,95]. However, the most used binders, such as polyvinylidene fluoride (PVDF), polyvinylidene fluoride-hexafluoropropylene (PVDF-HFP), and poly(methyl methacrylate) (PMMA), can only be dissolved in organic solvents, such as acetone,

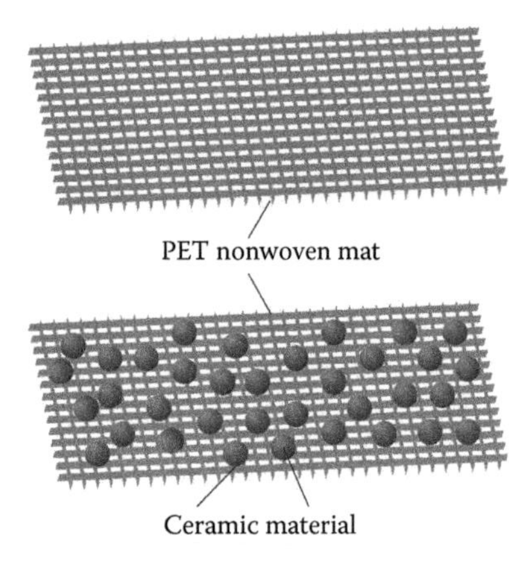

FIGURE 4.9 Structure of a ceramic separator.

dimethyl sulfoxide (DMSO), and tetrahydrofuran (THF), which make the coating process not eco-friendly [96,97]. Some water-based binders, i.e., carboxymethyl cellulose (CMC) and styrene butadiene rubber (SBR) mix binder, have been used to form the ceramic coating layer on the polyethylene (PE) separator with Al_2O_3 particles [98,99]. But, the modified separator cannot maintain integrity over 200°C. The separators are easily penetrated during the cells assembling process and the lithium dendrites formed in the process of charge–discharge, which rise the risk of internal shortcircuit [100]. Since thinner separators are more welcome to save more room for electrode materials to increase the energy density of cells, the puncture strength stands out to be a key mechanical property that affects the safety of LISC [101,102]. A water-based binder has been synthesized by grafting carboxyl groups onto cellulose diacetate. When the PE separator is coated by this binder and SiO_2 nanoparticles, the thermal shrinkage of the modified separator is observed to be almost 0% after exposure at 200°C for 30 min (Figure 4.10) [53]. The puncture strength significantly increased from 5.10 MPa (PE separator) to 7.64 MPa. The capacity retention of the cells assembled with modified separators after 100 cycles at 0.5 C increases from 73.3% (cells assembled with the PE separator) to 81.6%, owing to the excellent electrolyte uptake and the good compatibility with lithium electrode. Besides, the modified separator shows excellent surface stability after 100 cycles. Considering the above excellent properties, this composite separator shows high potential to be used in LISC with high power density and safety.

With the development of intelligent technology, a smart membrane separator has been developed for application in energy storage devices. Recently, a membrane, referred to as a smart membrane separator, that can regulate transmembrane ion transport as a function of its redox state has been designed (Figure 4.11), which is defined as a programmable ionic conductor that exhibits continuously varying ionic impedance due to an external stimulus and applied in an energy storage device [103].

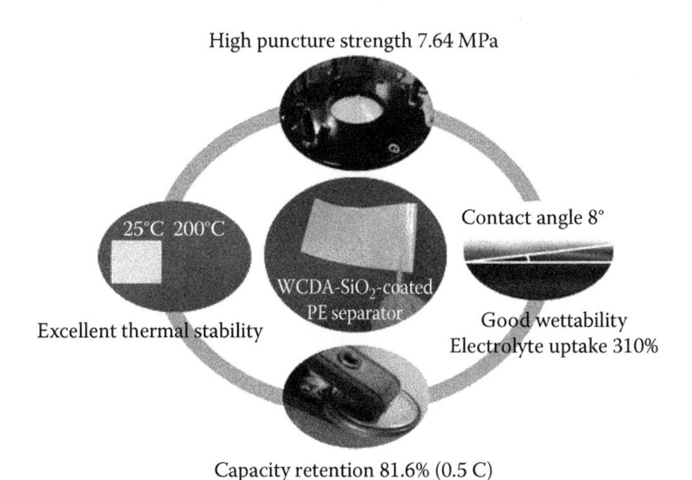

FIGURE 4.10 Water-based organic–inorganic hybrid coating for a high-performance separator [53]. Reprinted with permission from Ref. [53]. Copyright 2016, American Chemical Society.

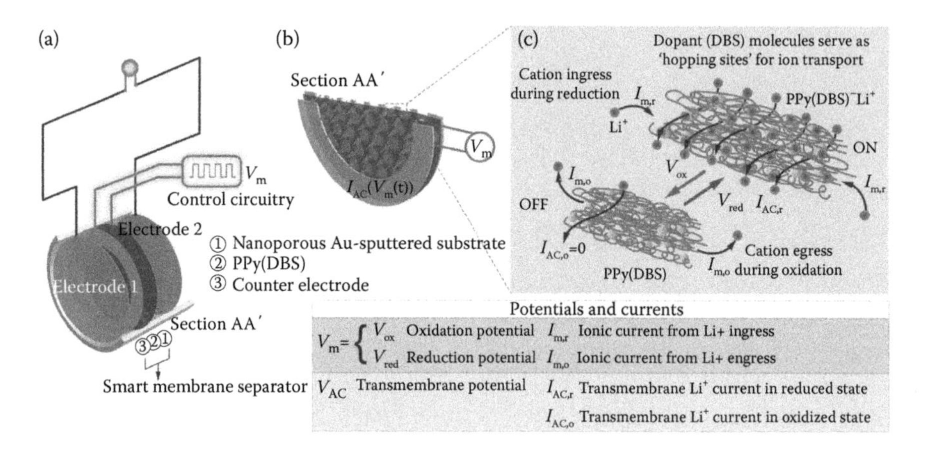

FIGURE 4.11 Structure and function of a smart membrane separator. (a) Implementation in an energy storage device uses double-layer electrodes (similar to supercapacitor) or liquid electrodes (similar to redox flow batteries) kept separated by the smart membrane separator. A control circuitry is connected to the membrane separator regulate the ionic impedance and ion transport between the electrodes. (b) PPy(DBS) is formed over a porous substrate spanning the pores. Electrical potential applied across the thickness of the PPy(DBS) using a counter electrode and the conductive layer between PPy(DBS) and porous substrate varies the redox state of PPy(DBS). (c) In the oxidized state, transmembrane impedance is high and there is minimal current across PPy(DBS). In the reduced state, cation ingress into PPy(DBS) enables ion transport across the membrane via hopping through dopant sites in PPy(DBS) [103]. Reprinted with permission from Ref. [103]. Copyright 2016, Royal Society of Chemistry.

It demonstrated that an applied electrical potential to a conducting polymermembrane alters the redox state and allows it to conduct ions across the membrane. This transmembrane ion transport across polypyrrole doped with dodecylbenzenesulfonate (PPy(DBS)) could be attributed to three plausible mechanisms as the membrane switches from oxidized to reduced state: (i) hopping pathways across the membrane, (ii) increase in equivalent pore size, and (iii) reversible wettability of PPy(DBS) [103,104]. The smart membrane separator when used in a supercapacitor or a hybrid battery provides additional control input to preserve the state of charge (SOC), and prolong the shelf life. Membrane separators are the last frontier in energy storage research, and interesting architectures have been proposed using thermally responsive polymers for mitigating thermal runaway [105,106].

The performance of polyimide (PI) nonwoven separator used in high-power lithium-ion battery is investigated to evaluate the feasibility of its industrialized application [107]. PI nonwoven separator has a fibrous membrane consisting of bead-free, uniformly dispersed thin fibers with an average diameter of about 491 nm. The PI separator with porosity higher than 90% and relatively high electrolyte uptake exhibits excellent thermal stability without obvious shrinkage at 500°C (Figure4.12) and shows a sufficient tensile strength of 11 MPa to meet the demand of cell assembly and usage. As shown in Figure 4.12, the degree of capacity fading of the PI-LIB as a function of the discharge rate is relatively smaller compared

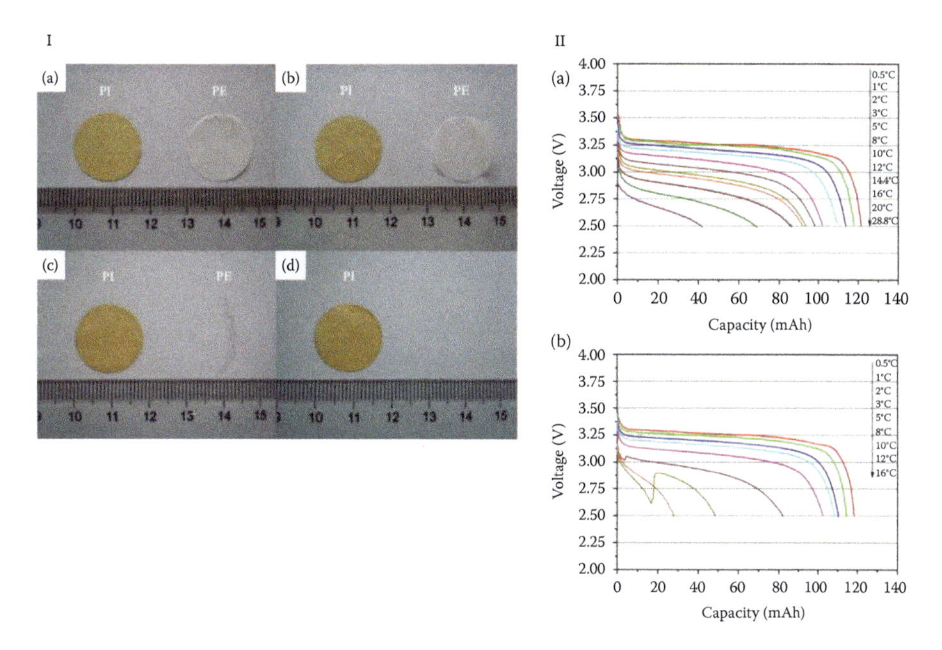

FIGURE 4.12 (I) Comparison of thermal stability between PI separator and the Celgard PE separator ((a) Before heating; (b) 150°C; (c) 250°C; (d) 500°C). (II) Discharge curves of the batteries at different discharge rates ((a) PI-LIB; (b) PE-LIB) [107]. Reprinted with permission from Ref. [107]. Copyright 2016, Elsevier.

with that of the PE-LIB. The PI-LIB shows no significant capacity fading and fluctuation, and the capacity retention ratio is 69.28% up to 16 C rate. At a super high discharge rate of 28.8 C, the capacity retention ratio is 33.6% and the discharge process remains stable. Thus, the PI electrospun separators have an overwhelming advantage over the Celgard PE separators in improving the high-rate capacity of lithium-ion batteries. This is mainly due to the considerably high porosity and tortuous 3D network structure of the electrospun separators that are beneficial to the absorption and retention of the electrolyte solution, and thereby facilitate the transference of electrolyte ions [108].

Cellulose can often be used as separator in aqueous electrolyte [109,110]. An asymmetric device was fabricated by using the Ni-based electrode as the cathode and the activated carbon on Ni foam as the anode, filter paper served as the separator. Figure 4.13a and b shows thet positive electrode has higher capacitance than that of negative electrode. Based on the specific capacitance from galvanostatic discharge curves, the symmetric supercapacitor displayed an energy density of 15.4 Wh/kg at power density of 750 W/kg (Figure 4.13c). The value of ESR is 1.4 Ω, further demonstrating the exceptional electrochemical performance of the device (Figure 4.13d) [111]. In the acidic electrolyte, filter paper can also be used as the separator. Two-electrode symmetric supercapacitors were fabricated using functionalized graphenes as anode and cathode materials in a 1 M H_2SO_4 electrolyte (Figure 4.14a) [112]. The rectangular outline of the CV curves (Figure 4.14b) shows redox peaks

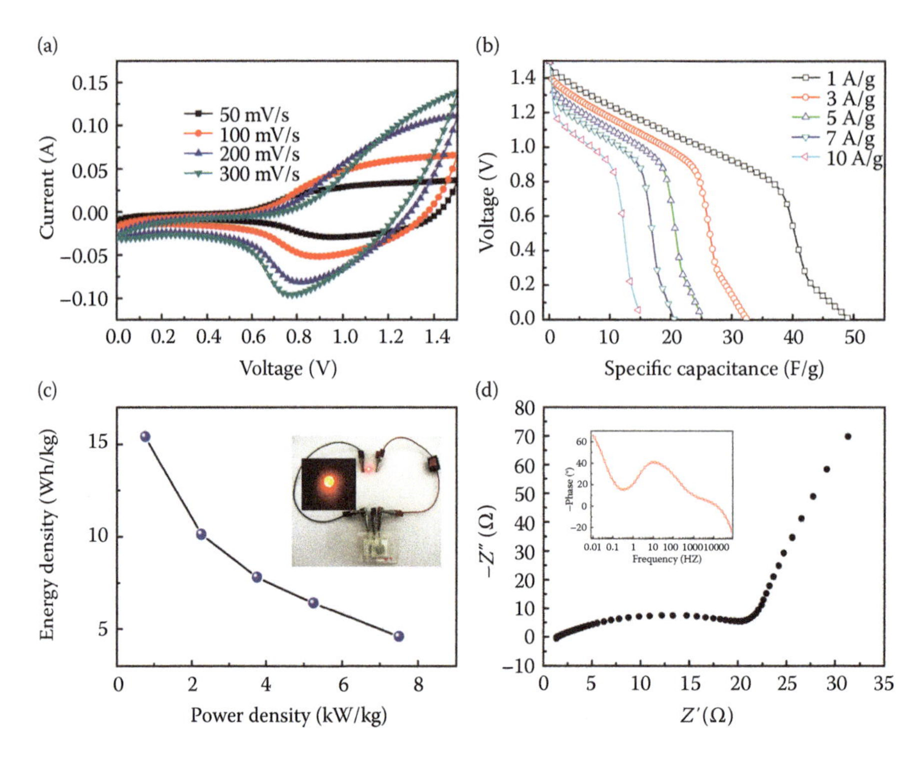

FIGURE 4.13 Electrochemical performances of activated Ni-based colloidal electrodes// activated carbon (AC) asymmetric device. (a) Cyclic voltammetry curves at various scan rates and (b) discharge curves of the asymmetric supercapacitor at various current densities. (c) Energy densities vs. power densities for asymmetric supercapacitor device. Inset shows a photograph of our asymmetric supercapacitor. (d) Nyquist and Bode plots [79]. Reprinted with permission from Ref. [79]. Copyright 2016, Royal Society of Chemistry.

with functionalized graphenes, indicating the presence of pseudocapacitance after the electrochemical oxidation processes. The symmetric supercapacitor displays excellent rate performance (Figure 4.14c and d) and cycling stability: after cycling at a current density of 20 A g^{-1} for 5000 cycles, the loss of specific capacitance was no more than 2% (Figure 4.14e). These results prove that cellulose films have a huge potential for practical applications.

4.6 COMPATIBILITY BETWEEN THE ELECTROLYTE AND THE SEPARATOR

The separator should quickly absorb and retain the electrolyte during operation. Poor wettability limits the performance of a cell by increasing the internal ionic resistance [113–115]. Good wettability shortens the electrolyte filling time during the cell assembly and extends the battery's life cycle under normal operating conditions [116,117]. When the wettability of the separator toward the electrolyte is poor, the pores in this separators are not completely filled with the liquid electrolyte, and thus, a high resistance will result

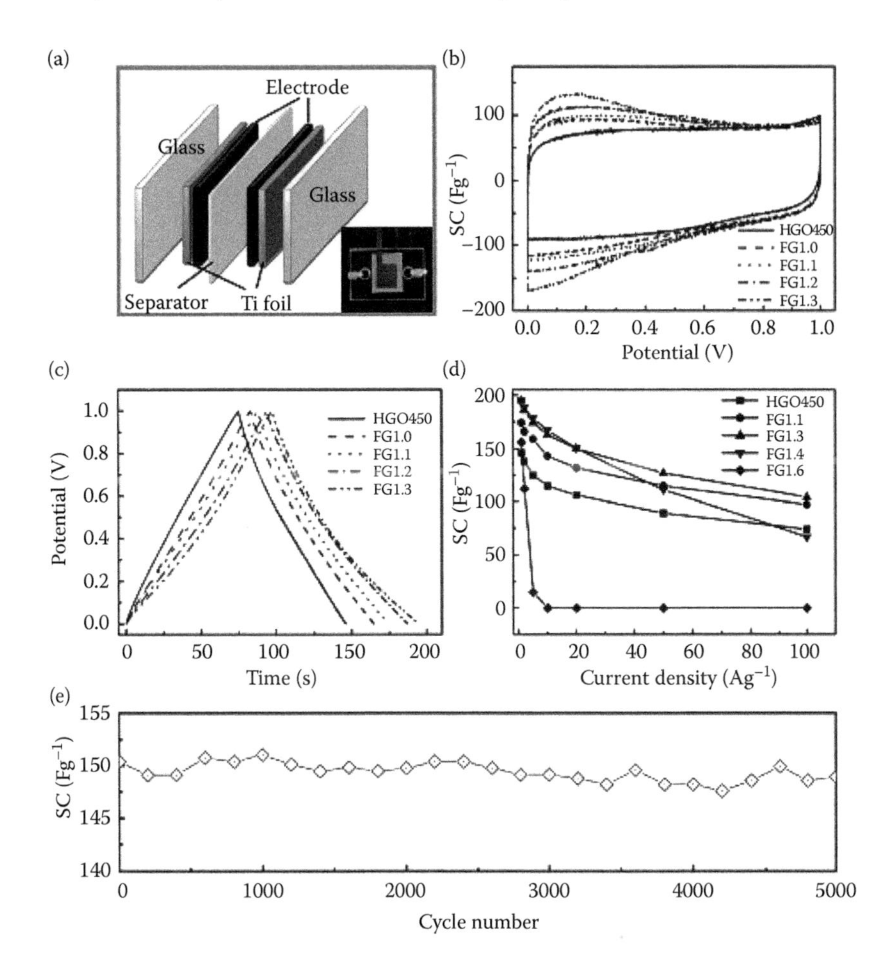

FIGURE 4.14 Supercapacitor tests of pristine graphenes (HGO450, reduced GO calcined at 450°C) and functionalized graphenes (FGs) in a symmetric two-electrode system: (a) illustration of the cell configuration for the supercapacitance test; (b) CV curves of HGO450 and FGs at a scan rate of 20 mV s^{-1} (c) galvanostatic charge–discharge curves of HGO450 and FGs at a current density of 1 A g^{-1}; (d) specific capacitances of HGO450 and FGs with varying charge–discharge rate; (e) cycling tests for the FG1.3 electrode at a current density of 20 A g^{-1} [112]. Reprinted with permission. Copyright 2013, WILEY-VCH Verlag GmbH & Co. KGaA, Weinheim.

from the blocked paths for Li^+-ion transportation [118] that is shown in Figure 4.15 [119]. The electrolyte uptake was calculated using the following equation:

$$\text{Electrolyte update} = \frac{M - M_0}{M_0} \times 100\%, \tag{4.8}$$

where M_0 and M represent the weights of the separator discs before and after adsorbing electrolyte. The viscosity and dielectric constant of solvents have great influence on separator wettability.

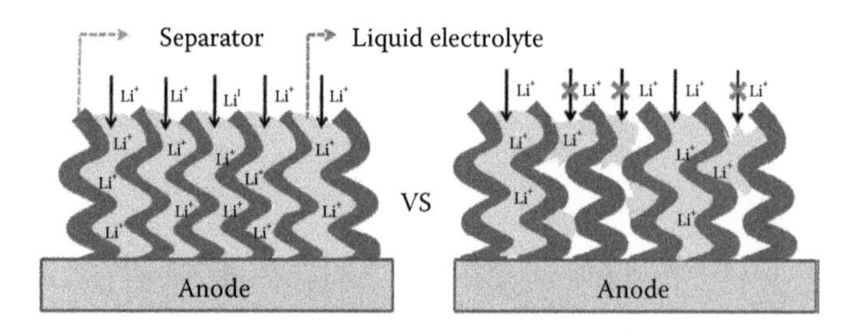

FIGURE 4.15 Schematic illustrations of the impacts of separator wettability toward electrolyte on lithium-ion transportation. The left picture represents good separator wettability, but the right one represents poor separator wettability toward electrolyte [119]. Reprinted with permission from Ref. [119]. Copyright 2016, Elsevier.

The wettability of a separator is important for the success of the electrolyte filling step during the cell assembly [120]. Cellulose has received much attention due to its excellent wettability, low processing cost, high porosity, good mechanical properties, lightweight, and sustainability [121–124]. Pure cellulose membranes with distinct structures and a variety of cellulose based composite membranes, e.g., cellulose/silica, cellulose/polydopamine have been manufactured and found to be comparable with or even better than polyolefin separators in terms of, e.g., electrolyte wettability, thermal stability, and cycling performance [125,126]. In Figure 4.16, the wettability of Cladophora cellulose (CC) and Solupor® are compared based on the electrolyte (i.e., LP40) spreading speed on these materials [127]. It is clearly seen that the electrolyte droplet on the CC surface spread out within seconds, while the one on the Solupor® separator exhibited almost the same size after 10 s. Since LP40 is a relatively polar electrolyte, this wettability difference can be ascribed to the larger hydrophilicity of the CC separator (and hence the higher hydrophobicity of the Solupor® separator). The latter results showed that a significantly lower contact angle was obtained for CC (i.e., 23°) than for Solupor® (i.e., 93°). The CC separators are expected to absorb electrolytes quickly, which should facilitate the electrolyte filling process. A $LiFePO_4$/Li cell containing a CC separator showed good cycling stability with 99.5% discharge capacity retention after 50 cycles at a rate of 0.2 C.

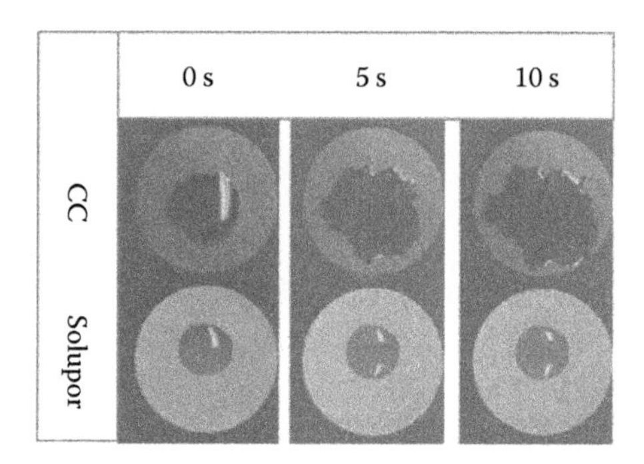

FIGURE 4.16 Electrolyte spreading wettability test of CC and Solupor® separators [127]. Reprinted with permission from Ref. [127]. Copyright 2016, Elsevier.

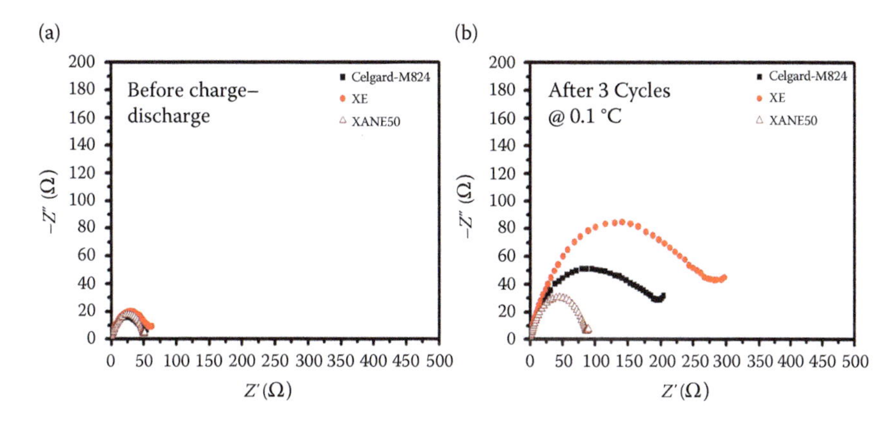

FIGURE 4.17 AC-impedance spectra with respect to cycle numbers for XANE50, XE membrane, and commercial separator cells: (a) before charge–discharge and (b) after 3 cycles charge–discharge at 0.1 C [128]. Reprinted with permission from Ref. [128]. Copyright 2014, American Chemical Society.

Polymer electrolytes, serving as both the electrolyte and the separator, can show good compatibility between the electrolyte and the separator. A polyacrylonitrile (PAN)-interpenetrating cross-linked polyoxyethylene (PEO) network (named XANE) was synthesized acting as the separator and as gel polymer electrolytes simultaneously. The porous XANE showed excellent electrolyte uptake amount (425 wt%), and electrolyte retention for XANE membrane, significantly higher than that of commercial separator (200 wt%) [128]. As shown in Figure 4.17, before charge–discharge cycling, the magnitudes of the interfacial resistances of the cell composed with XANE50, XE membrane, and a commercial separator were fairly similar (~50 Ω). However, after three charge–discharge cycles at 0.1 C, the resistance value of the commercial separator cell (190 Ω) was nearly double that of the XANE50 membrane cell (85 Ω). In addition, the resistance value of the XE membrane was rapidly increased to 275 Ω. Thus, we can conclude that the XANE50 membrane does not produce a marked increase in electrolyte–electrode interface resistance after charge–discharge cycling because of the better compatibility between the electrode and XANE membrane [129].

4.7 FAILURE MODE ANALYSIS AND MITIGATION STRATEGIES

Table 4.5 shows a comprehensive failure mode list of electrolyte and separator, the mode by which the failure is observed, the potential causes of the failure, whether the failure is brought on by progressive degradation (wearout) or abrupt overstress, the frequency of occurrence, the severity of failure, and the ease of detection of the failure mechanism [130–132]. If the integrity of the separator is compromised, the electrodes can make contact, resulting in an internal short-circuit that can cause heat generation and potentially lead to full thermal runaway [133]. Typically, electrode contact is due to thermal or mechanical damage to the separator. If metallic particles

TABLE 4.5
Failure Modes, Mechanisms, and Effects Analysis of Separator and Electrolyte

Component	Potential Failure Mode (s)	Potential Failure Mechanism (s)	Mechanism Type	Observed Effect	Potential Failure Causes	Likelihood of Occurrence
Separator	Hole in separator	Mechanical damage	Overstress	High heat generation due to joule heating, bloating of cell casing, drastic voltage reduction	Dendrite formation, external crushing of cell	Low
	Closing of separator pores	Thermally induced melting of separator	Overstress	Inability to charge or discharge battery	High internal cell temperature	Low
Electrolyte salt	Decrease in lithium salt concentration	Chemical reduction reaction and deposition	Wearout	Increased diffusion resistance	Chemical side reactions between lithium, electrodes, and solvent	High
Organic solvents	Gas generation and bloating of cell casing	Chemical decomposition of solvent	Overstress	Increased diffusion resistance, and may lead to thermal runaway	High external temperature, overcharging of the cell	Low
	Thickening of solid electrolyte interphase layer	Chemical reduction reaction and deposition	Wearout	Increased charge transfer resistance, reduction of capacity, reduction of power	Chemical side reactions between lithium, electrodes, and solvent	High

Source: Reprinted with permission from Ref. [132]. Copyright 2015, Elsevier.

are present in the battery and puncture the separator, internal shortcircuiting can occur. These particles can come from external contamination during assembly, lithium dendrite growth, and/or copper dissolution. Thermal and mechanical damage to the separator can be observed through an SEM. Additionally, the melted separator can be seen optically during a cell disassembly procedure and provide some insight into the initiation point of the failure [134]. As a safety mechanism, the pores of the separator are designed to close to prevent shortcircuiting [135]. However, if the battery continues to heat up, the separator could further melt and allow an internal short circuit. Finally, the separator is usually wider than the electrodes to prevent short circuiting at the edges. However, separator shrinkage when exposed to electrolyte or high temperatures could cause the electrode edges to touch, possibly initiating a thermal runaway [136]. Proper design and selection of the separator can aid in preventing thermal runaway, but proper manufacturing and screening procedures are required to prevent contaminated cells from entering the supply chain.

Degradation mechanisms of the electrolyte include the following [137]: (1) parasitic (and most likely irreversible) reactions between electrolyte components and charged electrodes; (2) the effect of trace moisture, which is closely intertwined with dissolution of transition metals from cathodes and inter-reaction among electrolyte components; (3) the dissolution of active electrode materials in electrolytes, which is aided by the presence of acidic impurities (HF generated during hydrolysis of PF_6^- by trace moisture); (4) the inter-reactions of bulk electrolyte components, such as the trans-esterification of carbonates, and similar nucleophilic reactions between cyclic and acyclic carbonates leading to the formation of undesirable alkyldicarbonates; (5) dissolution of interphasial components and possible "electrochemical dialogs" between the cathode and anode; (6) corrosion of current collectors by electrolyte components at extreme potentials; and (7) thermal effects, which include both low (subzero) and high (>40°C) temperatures. Table 4.6 shows the failure mechanism of aqueous electrolyte in Li_2SO_4 solution [138]. The oxidation and reduction of water is the major failure mechanism of aqueous electrolyte. The development of aqueous electrolyte with high overpotential of H_2 and O_2 evolution is necessary.

If operated under extreme temperature or voltage conditions, the electrolyte can decompose, cause gas generation, and lead to thermal runaway [150]. Typically, electrolyte decomposition occurs at temperatures above 200°C, at which point thermal runaway becomes likely [151]. Alternately, researchers have focused on heat generation and heat dissipation rate when modeling thermal abuse that would lead to electrolyte decomposition [151]. Thermal abuse models evaluate the contributions of different sources of heat generation, including short circuits and chemical reactions, to identify a timeline for the initiation of thermal runaway.

To mitigate the effect of electrolyte degradation on cell performance, new electrolyte components were developed in attempts to slow the above-mentioned processes. Replacement of $LiPF_6$ by lithium salts of higher thermal stability was also attempted, where LiFAP, LiBOB, and other borate variations were identified to help mitigate the parasitic reactions at elevated temperatures [152]. The presence of a more electrophilic additive such as VC would effectively suppress the formation of alkyl dicarbonates, leading to high stability. More efforts were made to reduce the inflammability of carbonate-based electrolytes, so that catastrophic chain

TABLE 4.6
Summary of Possible Parasitic Reactions in Activated Carbon Based EDLC Cells with 1 mol L^{-1} Li_2SO_4 Electrolyte

Process	*U*(V)	*E*(V vs. SHE)	Reactions	Ref.
Carbon surface group oxidation	~0.6	0.25 (~0.6, pH = 1)	$(RCO)_2O - 4e^- \rightarrow 2R + CO\uparrow + CO_2\uparrow$ $R_1COCOR_2O - 4e^- \rightarrow R_1 + R_2 + 2CO\uparrow$	139
Carbon surface group reduction	n.a.	n.a. (~0.1, pH = 10)	$R_1COR_2 + 2e^- + 2H^+ \rightarrow R_1CHOHR_2$ $R_1COCOR_2 + 4e^- + 4H^+ \rightarrow R_1(CHOH)_2R_2$	140
H_2 chemisorption	n.a.	n.a.(<0.35, pH = 7)	$AC + xH_2O + xe^- \rightarrow ACH_x + XOH^-$	141, 142
H_2 evolution	>1.6	< −0.79	$CH_{ad} + H_2O + e^- \rightarrow H_2\uparrow + OH^- + C$ $CH_{ad} + CH_{ad} \rightarrow H_2\uparrow + C$	143
O_2 evolution	None	None	$2H_2O - 4e^- \rightarrow O_2\uparrow + 4H^+$	144
Carbon corrosion	>1.6	>0.81	$H_2O - ne^- \rightarrow OH^- + H^+(n = 1),\quad O^- + 2H^+(n = 2)$ $C + O^-/OH^- \rightarrow C = O, C - OH_{ads} \rightarrow CO(+H^+)\uparrow$ $C + 2H_2O - 4e^- \rightarrow CO/CO_2\uparrow + 4H^+$	145
Oxidation of hydrogen	n.a.	n.a. (~0.2, pH =0)	$2CH_{ad} - 2e^- \rightarrow 2C + 2H^+$ $H_2 - 2e^- \rightarrow 2H^+$	146
Trace O_2 reduction	~0.4	~0.1	$O_2 + H_2O + 2e^- \rightarrow HO_2^- + OH^-$ $O_2 + 4H_2O + 4e^- \rightarrow 4OH^-$	147
CO oxidation	n.a.	n.a.	$CO + O/O_2 \rightarrow CO_2\uparrow$	148
CO_2 chemical absorption	n.a.	n.a.	$CO_2 + 2OH^- \rightarrow H_2O + CO_3^{2-}$ $2Li^+ + CO_3^{2-} \rightarrow Li_2CO_3$	149

U and *E* represent the cell voltage and electrode potential, respectively.
Source: Reprinted with permission from Ref. [138]. Copyright 2016, Elsevier.

reactions, mainly contributed by electrolytes, could be avoided in case of accidental abuses such as overcharging, mechanical shorting, and overheating. For example, phosphorus-containing compounds were explored as potential candidates for nonflammable electrolytes [153]. Various new solvents, additives, or salts have been developed, targeting the interphasial stabilities between specific electrolyte/anode couples. The reactivity between the electrolyte and the delithiated cathode is usually the main source of heat generation leading to catastrophic thermal runaway. Efforts to mitigate this process by developing electrolyte components proved more challenging than the corresponding work on the anode side, because the electrochemical oxidation of electrolyte solvents or any other organic components (binders, separators) is hardly affected by changes in electrolyte composition or interphasial chemistry [154]. Another major source of degradation that involves electrolyte components is the corrosion of electrode substrates, particularly the anodic dissolution of cathode substrates at high potentials [155]. The passivation of aluminum relies more on the salt (or salt anion) species rather than solvents [156], differing from the interphasial chemistry on graphite anodes. PF_6^- is one of the anions that can passivate aluminum at both high potentials and elevated temperatures. Most imide salts would inherit the corrosive nature of LiTFSI toward aluminum [157].

4.8 CHALLENGES AND PERSPECTIVES

High power and energy density of LISCs are being developed for use in hybrid electric vehicles for fuel efficiency improvement and greenhouse gas emissions reduction. Due to the high power density of LISCs, specific electrolyte and separator are needed. Now, the commercial electrolyte and separator for lithium-ion batteries are often used in LISCs. Significant efforts are being made to construct high-performance LISCs through the design of the electrolyte and separator as well as the control of the solid-electrolyte interface. The nanoporous separators facilitate fast, liquid-like ion transport because of liquid electrolytes trapped in the pores. The rigid ceramic or solid polymeric framework of the separators brings high mechanical modulus on the other hand. Electrolyte design should markedly increase the rate of ion transport across the interface and, at the same time, protect the electrode materials and current collector from side reactions with electrolyte solvent. It is also important to significantly increase the electrode lifetime using novel additives in the electrolyte, even in the absence of a separator.

The interface design of electrode materials–electrolyte, electrolyte–separator, electrode materials–separator is important for constructing high power, high stability, and high safety LISC devices. The reduction of interface resistance, the increase of interface wettability and stability are important issues. Polymer and solid electrolytes being served as both the electrolyte and the separator in one are designed to overcome above issues. The proper use of chemical reactions between electrode materials and electrolytes can produce new interface between electrode materials and the electrolyte, which can increase electrochemical performance of LISCs. For example, high activated colloidal ionic supercapacitor electrodes have been synthesized by using in situ chemical and electrochemical activation reactions, which can show multiple-electron redox reaction compared with traditional electrode.

To gain insight into the chemical reactions involving the multiple components (electrodes, electrolytes, separator, interfaces) in the electrochemical cells and to determine the failure mechanism, advanced characterization tools over multiple length scales are necessary. Fundamental research and development is needed to enable technological breakthroughs and increase market penetration of LISC. In situ and real-time characterization techniques have played an important role in increasing the fundamental understanding on the structure and dynamics of different components in the devices, including the electrodes, separators, and liquid and solid electrolytes.

4.9 CHAPTER SUMMARY

With the objective of improving the energy density, LISCs were developed by introducing battery and supercapacitor materials. As two key components of LISCs, electrolyte and separator, can determine the voltage range, power performance, resistance, and safety of LISCs. Advanced organic electrolytes, ionic liquids, and redox electrolytes can create higher electrochemical capacitors due to the increase of voltage window, stability of devices, and specific capacitance of LISCs. In addition to porous polymer separators, polymer electrolytes and solid electrolytes, serving as both the separator and the electrolyte, have been provided to increase the electrochemical performance and safety of LISCs. With the full understanding of the failure mechanisms of electrolytes and separators, novel electrolytes and separators can be designed. LISCs often show different requirements compared with lithium-ion battery and traditional supercapacitor; therefore, the design of novel electrolytes and separators needs to be tailored to address this issue.

ACKNOWLEDGMENTS

Financial support from the National Natural Science Foundation of China (Grant Nos. 51125009, 91434118, 21601176), the National Natural Science Foundation for Creative Research Group (Grant No. 21521092), the External Cooperation Program of BIC, Chinese Academy of Sciences (Grant No. 121522KYS820150009), the Hundred Talents Program of the Chinese Academy of Sciences, and Jilin Provincial Science and Technology Development Program of China (Grant No. 20160520002JH) is acknowledged.

REFERENCES

1. Manthiram, A., Fu, Y., and Su, Y. 2013. In charge of the world: electrochemical energy storage. *J. Phys. Chem. Lett.* 4: 1295–7.
2. Simon, P., and Gogotsi, Y. 2008. Materials for electrochemical capacitors. *Nat. Mater.* 7: 845–854.
3. Conway, B. E. *Electrochemical Supercapacitors: Scientific Fundamentals and Technological Applications*, Kluwer-Academic, New York, 1999.
4. Naoi, K. 2013. Evolution of energy storage on the platform of supercapacitors. *Electrochemistry* 81: 775–6.
5. Naoi, K., Ishimoto, S., Miyamoto, J. I., and Naoi, W. 2012. Second generation 'nanohybrid supercapacitor': evolution of capacitive energy storage devices. *Energy Environ. Sci.* 5: 9363–73.
6. Yoo, H. D., Markevich, E., Salitra, G., Sharon, D., and Aurbach, D. 2014. On the challenge of developing advanced technologies for electrochemical energy storage and conversion. *Mater. Today* 17: 110–21.
7. Chen, K., Song, S., Liu, F., and Xue, D. 2015. Structural design of graphene for use in electrochemical energy storage devices. *Chem. Soc. Rev.* 44: 6230–57.
8. Chen, K., Pan, W., and Xue, D. 2016. Phase transformation of Ce^{3+}-doped MnO_2 for pseudocapacitive electrode materials. *J. Phys. Chem. C* 120: 20077–81.
9. Chen, K., and Xue, D. 2016. Colloidal supercapacitor electrode materials. *Mater. Res. Bull.* 83: 201–6.
10. Chen, K., and Xue, D. 2016. Materials chemistry toward electrochemical energy storage. *J. Mater. Chem. A* 4: 7522–37.
11. Chen, K., and Xue, D. 2015. Beyond graphene: materials chemistry toward high performance inorganic functional materials. *J. Mater. Chem. A* 3: 2441–53.
12. Aravindan, V., Gnanaraj, J., Lee, Y., and Madhavi, S. 2014. Insertion-type electrodes for nonaqueous li-ion capacitors. *Chem. Rev.* 114: 11619–35.

13. Thangavel, R., Kaliyappan, K., Kang, K., Sun, X. L., and Lee, Y. S. 2016. Going beyond lithium hybrid capacitors: proposing a new high-performing sodium hybrid capacitor system for next-generation hybrid vehicles made with bio-inspired activated carbon. *Adv. Energy Mater.* 6: 1502199.
14. Yang, M., Zhong, Y. R., Ren, J. J., Zhou, X. L., Wei J. P., and Zhou, Z. 2015. Electrochemical capacitors: fabrication of high-power li-ion hybrid supercapacitors by enhancing the exterior surface charge storage. *Adv. Energy Mater.* 5(17): 2416–20.
15 Wang, H. W., Zhang, Y., Ang, H. X., Zhan, Y. Q., Tan, H. T., Zhang, Y. F., Guo, Y. Y., Franklin, J. B., Wu, X. L., Srinivasan, M., Fan, H. J., and Yan, Q. Y. 2016. A high-energy lithium-ion capacitor by integration of a 3D interconnected titanium carbide nanoparticle chain anode with a pyridine-derived porous nitrogen-doped carbon cathode. *Adv. Funct. Mater.* 26(18): 3082–93.
16. Cairns, E. J., and Albertus, P. 2010. Batteries for electric and hybrid-electric vehicles. *Annu. Rev. Chem. Biomol. Eng.* 1: 299–320.
17. Xu, K. 2014. Electrolytes and interphases in Li-ion batteries and beyond. *Chem. Rev.* 114(23): 11503–618.
18. Frackowiak, E., Abbas, Q., and Beguin, F. 2013. Carbon/carbon supercapacitors. *J. Energy Chem.* 22(2): 226–40.
19. Fic, K., Meller, M., and Frackowiak, E. 2014. Strategies for enhancing the performance of carbon/carbon supercapacitors in aqueous electrolytes. *Electrochim. Acta* 128(5): 210–7.
20. Béguin, F., Presser, V., Balducc, A., and Frackowiak, E. 2014. Carbons and electrolytes for advanced supercapacitors. *Adv. Mater.* 26(14): 2219–51.
21. Arora, P., and Zhang, Z. 2014. Battery separators. *Chem. Rev.* 104 (10): 4419–62.
22. Kritzer, P., and Cook, J. A. 2007. Nonwovens as separators for alkaline batteries an overview. *J. Electrochem. Soc.* 154 (5): A481–94.
23. Hwang, H. J., Chi, W. S., Kwon, O., Lee, J. G., Kim, J. H., and Shul, Y. G. 2016. Selective ion transporting polymerized ionic liquid membrane separator for enhancing cycle stability and durability in secondary zinc air battery systems. *ACS Appl. Mater. Interfaces* doi: 10.1021/acsami.6b07841.
24. Xu, K. 2004. Nonaqueous liquid electrolytes for lithium-based rechargeable batteries. *Chem. Rev.* 104(10): 4303–417.
25. Dunn, B., Kamath, H., and Tarascon, J. M. 2011. Electrical energy storage for the grid: a battery of choices. *Science* 334: 928–35.
26. Xu, K., and Cresce, A. J. 2011. Interfacing electrolytes with electrodes in Li ion batteries. *J. Mater. Chem.* 21: 9849–64.
27. Marom, R., Amalraj, S. F., Leifer, N., Jacob, D., and Aurbach, D. J. 2011. A review of advanced and practical lithium battery materials. *J. Mater. Chem.* 21: 9938–54.
28. Chen, K., Yin, S., and Xue, D. 2015. A binary A_xB_{1-x} ionic alkaline pseudocapacitor system involving manganese, iron, cobalt, and nickel: formation of electroactive colloids via in situ electric field assisted coprecipitation. *Nanoscale* 7: 1161–6.
29. Acerce, M., Voiry, D., and Chhowalla, M. 2015. Metallic 1T phase MoS_2 nanosheets as supercapacitor electrode materials. *Nat. Nanotechnol.* 10: 313–8.
30. Wang, G., Zhang, L., and Zhang, J. 2012. A review of electrode materials for electrochemical supercapacitors. *Chem. Soc. Rev.* 41(2): 797–828.
31. Yan, J., Wang, Q., Wei, T., and Fan, Z. J. 2014. Recent advances in design and fabrication of electrochemical supercapacitors with high energy densities. *Adv. Energy Mater.* 4: 1300816.
32. Gao, Q., Demarconnay, L., Raymundo-Piñero, E., and Béguin, F. 2012. Exploring the large voltage range of carbon/carbon supercapacitors in aqueous lithium sulfate electrolyte. *Energy Environ. Sci.* 5(11): 9611–7.

33. Fic, K., Lota, G., Meller, M., and Frackowiak, E. 2012. Novel insight into neutral medium as electrolyte for high-voltage supercapacitors. *Energy Environ. Sci.* 5(2): 5842–50.
34. Liu, F., Song, S., Xue, D., and Zhang, H. 2012. Folded structured graphene paper for high performance electrode materials. *Adv. Mater.* 24: 1089–94.
35. Chen, K., Yang, Y., Li, K., Ma, Z., Zhou, Y., and Xue, D. 2014. $CoCl_2$ designed as excellent pseudocapacitor electrode materials. *ACS Sustainable Chem. Eng.* 2: 440–4.
36. Chen, L. et al. 2016. Water-mediated cation intercalation of open-framework indium hexacyanoferrate with high voltage and fast kinetics. *Nat. Commun.* 7: 11982.
37. Hertzberg, B., Huang, A., Hsieh, A., Chamoun, M., Davies, G., Seo, J. K., Zhong, Z., Croft, M., Erdonmez, C., Meng, Y. S., and Steingart, D. 2016. Effect of multiple cation electrolyte mixtures on rechargeable Zn MnO_2 alkaline battery. *Chem. Mater.* 28(13): 4536–45.
38. Chen, K. F., Li, G., and Xue, D. 2016. Architecture engineering of supercapacitor electrode materials. *Funct. Mater. Lett.* 9(1): 1640001.
39. Chen, K., and Xue, D. 2015. Rare earth and transitional metal colloidal supercapacitors. *Sci. China Technol. Sci.* 58(11): 1768–78.
40. Lota, G., and Frackowiak, E. 2009. Striking capacitance of carbon/iodide interface. *Electrochem. Commun.* 11: 87–90.
41. Roldán, S., Blanco, C., Granda, M., Menéndez R., and Santamaría, R. 2011. Towards a further generation of high-energy carbon-based capacitors by using redox-active electrolytes. *Angew. Chem. Int. Ed.* 50(7): 1699–701.
42. Frackowiak, E., Fic, K., Meller, M., and Lota G. 2012. Electrochemistry serving people and nature: high-energy ecocapacitors based on redox-active electrolytes. *ChemSusChem* 2012(5): 1181–5.
43. Béguin, F., Presser, V., Balducci, A., and Frackowiak, E. 2014. Carbons and electrolytes for advanced supercapacitors. *Adv. Mater.* 26: 2219–51.
44. Chiba, K. 2008. Electrolytic solution for electric double layer capacitor and electric double layer capacitor. US Patent Office, US 20060274475 A1.
45. Chiba, K., Ueda, T., Yamaguchi, Y., Oki, Y., Saiki, F., and Naoi, K. 2011. Electrolyte systems for high withstand voltage and durability I. linear sulfones for electric double-layer capacitors. *J. Electrochem. Soc.* 158(12): A1320–7.
46. Chiba, K., Ueda, T., Yamaguchi, Y., Oki, Y., Shimodate, F., and Naoi, K. 2011. Electrolyte systems for high withstand voltage and durability I. linear sulfones for electric double-layer capacitors. *J. Electrochem. Soc.* 158: A872–82.
47. Naoi, K. 2010. 'Nanohybrid capacitor': the next generation electrochemical capacitors. *Fuel Cells* 10(5): 825–33.
48. Ruch, P. W., Cericola, D., Foelske, A., Kötz, R., and Wokaun, A. 2010. A comparison of the aging of electrochemical double layer capacitors with acetonitrile and propylene carbonate-based electrolytes at elevated voltages. *Electrochim. Acta* 55(7): 2352–57.
49. Brandt, A., Balducci, A., Rodehorst, U., Menne, S., Winter, M., and Bhaskara, A. 2014. Investigations about the use and the degradation mechanism of LiNi0.5Mn1.5O4 in a high power LIC. *J. Electrochem. Soc.* 161(6): A1139–43.
50. Wang, F., Gan, C. L., Yuan, and X. Y. 2016. Industrial progress of nonaqueous liquid electrolyte for lithium-ion batteries. *Energy Storage Sci. Technol.* 5(1): 1–8.
51 Hall, P. J., Mirzaeian, M., Fletcher, S. I., Sillars, F. B., Rennie, A. J. R., Shitta-Bey, G. O., Wilson, G., Cruden, A., and Carter, R. 2010. Energy storage in electrochemical capacitors: designing functional materials to improve performance. *Energy Environ. Sci.* 3(9): 1238–51.
52. Harnisch, F., Schröder, U., and Scholz, F. 2008. The suitability of monopolar and bipolar ion exchange membranes as separators for biological fuel cells. *Environ. Sci. Technol.* 42: 1740–46.

53. Chen, W. J., Shi, L. Y., Zhou, H. L., Zhu, J. F., Wang, Z. Y., Mao, X. F., Chi, M. M., Sun, L. L., and Yuan, S. 2016. Water-based organic–inorganic hybrid coating for a high-performance separator. *ACS Sustainable Chem. Eng.* 4: 3794–02.
54. Huang, X. S. 2011. Separator technologies for lithium-ion batteries. *J. Solid State Electrochem.* 15: 649–62.
55. Kirchhöfer M., Zamory J. V., Paillard E., and Passerini S. 2014. Separators for Li-Ion and Li-metal battery including ionic liquid based electrolytes based on the $TFSI^-$ and FSI^- anions. *Int. J. Mol. Sci.* 15: 14868–90.
56. Miao, Y. E., Yan, J. J., Huang, Y. P., Fan, W., and Liu, T. X. 2015. Electrospun polymer nanofiber membrane electrodes and an electrolyte for highly flexible and foldable all-solid-state supercapacitors. *RSC Adv.* 5: 26189–96.
57. Hsu, L., and Sheibley, D. 1982. Inexpensive cross-linked polymeric separators made from water-soluble polymers. *J. Electrochem. Soc.* 129(2): 251–4.
58. Takahashi, K., Tokuno, K., and Kihira, H. 2010. Titanium Material Having Low Contact Resistance for Use in Separator for Solid Polymer-Type Fuel Cell and Process for Producing the Titanium Material. EP 2 337 135 A1.
59. AndoU, N., and Takeuchi, M. 1998. Electrical resistivity of the polymer layers with polymer grafted carbon blacks. *Thin Solid Films* 334: 182–6.
60. Zhang, H., Lin, C. E., Zhou, M. Y., John, A. E., and Zhu, B.-K. 2016. High thermal resistance polyimide separators prepared via soluble precusor and non-solvent induced phase separation process for lithium ion batteries. *Electrochim. Acta* 187: 125–33.
61. Madsen, B. 2011. Centrifugal Separator. *US* 7981818B2.
62. Zhang, S. S. 2007. A review on the separators of liquid electrolyte Li-ion batteries. *J. Power Sources* 164(1): 351–64.
63. Liu, H. 2015. Research of Diaphragms and Electrolyte for the Super Capacitor. Ji Lin: Jilin University.
64. Venugopal, G., Moore, J., Howard, J., and Pendalwar, S. 1999. Characterization of microporous separators for lithium-ion batteries. *J. Power Sources* 77: 34–41.
65. Yu, L. H., Jin, Y., and Lin, Y. S. 2016. Ceramic coated polypropylene separators for lithium-ion batteries with improved safety: effects of high melting point organic binder. *RSC Adv.* 6: 40002–9.
66. Jeon, H., Jin, S. Y., Park, W. H., Lee, H. K., Kim, H. T., Ryou, M. H., and Lee, Y. M. 2016. Plasma-assisted water-based Al_2O_3 ceramic coating for polyethylene-based microporous separators for lithium metal secondary batteries. *Electrochim. Acta* 212: 649–56.
67. Wu, F., Chen, N., Chen, R. J., Zhu, Q. Z., Qian, Ji., and Li, L. 2016. "Liquid-in-solid" and "solid-in-liquid" electrolytes with high rate capacity and long cycling life for lithium-ion batteries. *Chem. Mater.* 28(3): 848–56.
68. Chen, K., Donahoe, A. C., Noh, Y., Li, K. Y., Komarneni, S., and Xue, D. 2014. Conventional- and microwave-hydrothermal synthesis of $LiMn_2O_4$: effect of synthesis on electrochemical energy storage performances. *Ceram. Int.* 40(2): 3155–63.
69. Wang, H. X., Zhang, W., Chen, H., and Zheng, W. T. 2015. Towards unlocking high-performance of supercapacitors: from layered transition-metal hydroxide electrode to redox electrolyte. *Sci. China: Technol. Sci.* 58: 1779–98.
70. Yu, H. J., Wu, J. H., Fan, L. Q., Hao, S. C., Lin, J. M., and Huang, M. L. 2014. An efficient redox-mediated organic electrolyte for high-energy supercapacitor. *J. Power Sources* 248: 1123–26.
71. Senthilkumar, S. T., Selvan, R. K., Ulaganathan, M., and Melo, J. 2014. Fabrication of Bi_2O_3||AC asymmetric supercapacitor with redox additive aqueous electrolyte and its improved electrochemical performances. *Electrochim. Acta* 115: 518–24.
72. Zhao, C., Zheng, W., Wang, X. et al. 2013. Ultrahigh capacitive performance from both $Co(OH)_2$/graphene electrode and $K_3Fe(CN)_6$ electrolyte. *Sci. Rep.* 3: 2986.

73. Yu, H., Wu, J., Fan, L. et al. 2012. A novel redox-mediated gel polymer electrolyte for high-performance supercapacitor. *J. Power Sources* 198: 402–7.
74. Chen, K., Liu, F., Xue, D., and Komarneni, S. 2015. Carbon with ultrahigh capacitance when graphene paper meets $K_3Fe(CN)_6$. *Nanoscale* 7(2): 432–9.
75. Chen, K., Song, S. Y., and Xue, D. 2014. An ionic aqueous pseudocapacitor system: electroactive ions in both a salt electrode and redox electrolyte. *RSC Adv.* 4(44): 23338–43.
76. Chen, X., Chen, K., Wang, H., and Xue, D. 2015. A colloidal pseudocapacitor: direct use of $Fe(NO_3)_3$ in electrode can lead to a high performance alkaline supercapacitor system. *J. Colloid Interface Sci.* 444: 49–57.
77. Chen, K., and Xue, D. 2014. Ionic supercapacitor electrode materials: a system-level design of electrode and electrolyte for transforming ions into colloids. *Colloids Interface Sci. Commun.* 1: 39–42.
78. Chen, K., and Xue, D. 2014. Formation of electroactive colloids via in-situ coprecipitation under electric field: erbium chloride alkaline aqueous pseudocapacitor. *J. Colloid Interface Sci.* 430: 265–71.
79. Chen, K., and Xue, D. 2016. In situ electrochemical activation of Ni-based colloids from $NiCl_2$ electrode and their advanced energy storage performance. *Nanoscale* 8: 17090–5.
80. Chen, X., Chen, K., Wang, H., Song, S., and Xue, D. 2014. Crystallization of Fe^{3+} in an alkaline aqueous pseudocapacitor system. *CrystEngComm* 16: 6707–15.
81. Chen, K., and Xue, D. 2014. $YbCl_3$ electrode in alkaline aqueous electrolyte with high pseudocapacitance. *J. Colloid Interface Sci.* 424: 84–9.
82. Lu, P., Liu, F., Xue, D., Yang, H., and Liu, Y. 2012. Phase selective route to $Ni(OH)_2$ with enhanced supercapacitance: performance dependent hydrolysis of $Ni(Ac)_2$ at hydrothermal conditions. *Electrochim. Acta* 78: 1–10.
83. Zhu, S., Jia, J., Wang, T., Zhao, D., Yang, J., Dong, F., Shang, Z., and Zhang, Y. X. 2015. Rational design of octahedron and nanowire CeO_2@MnO_2 core-shell heterostructures with outstanding rate capability for asymmetric supercapacitors. *Chem. Commun.* 51(80): 14840–3.
84. Forse, A. C., Merlet, C., Griffin, J. M., and Grey, C. P. 2016. New perspectives on the charging mechanisms of supercapacitors. *J. Am. Chem. Soc.* 138: 5731–44.
85. El-Kady, M. F., Shao, Y., and Kaner, R. B. 2016. Graphene for batteries, supercapacitors and beyond. *Nat. Rev. Mater.* 1: 16033.
86. Chen, K., and Xue, D. 2014. Ex situ identification of the Cu^+ long-range diffusion path of a Cu-based anode for lithium ion batteries. *Phys. Chem. Chem. Phys.* 16(23): 11168–72.
87. Saito, Y., Morimura, W., Kuratani, R., and Nishikawa, S. 2015. Ion transport in separator membranes of lithium secondary batteries. *J. Phys. Chem. C* 119(9): 4702–08.
88. Xia, M., Liu, Q. Z., Zhou, Z., Tao, Y. F., Li, M. F., Liu, K., Wu, Z. H., and Wang, D. 2014. A novel hierarchically structured and highly hydrophilic poly(vinyl alcohol-co-ethylene)/poly(ethylene terephthalate) nanoporous membrane for lithium-ion battery separator. *J. Power Sources* 266: 29–35.
89. Yang, C. R., Jia, Z. D., Guan, Z. C., and Wang, L. M. 2009. Polyvinylidene fluoride-membrane by novel electrospinning systemfor separator of Li-ion batteries. *J. Power Sources* 189(1): 716–20.
90. Hennige, V., Hying, C., and Horpel, G. 2008. Ceramic Separator for Electrochemcial Cells with Improved Conductivity. US Patent Office, US 20080248381 A1.
91. Park, J. H., Cho, J. H., Park, W., Ryoo, D., Yoon, S. J., Kim, J. H., Jeong, Y. U., and Lee, S. Y. 2010. Close-packed SiO_2/poly(methyl methacrylate) binary nanoparticles-coated polyethylene separators for lithium-ion batteries. *J. Power Sources* 195(24): 8306–10.

92. Wang, H., Wu, J., Cai, C., Guo, J., Fan, H., Zhu, C., Dong, H., Zhao, N., and Xu, J. 2014. Mussel inspired modification of polypropylene separators by catechol/polyamine for Li-ion batteries. *ACS Appl. Mater. Interfaces* 6(8): 5602–8.
93. Jeong, H. S., Hong, S. C., and Lee, S. Y. 2010. Effect of microporous structure on thermal shrinkage and electrochemical performance of Al_2O_3/poly(vinylidene fluoride-hexafluoropropylene) composite separators for lithium-ion batteries. *J. Membr. Sci.* 364(1): 177–82.
94. Zhang, Z., Lai, Y., Zhang, Z., Zhang, K., and Li, J. 2014. Al_2O_3-coated porous separator for enhanced electrochemical performance of lithium sulfur batteries. *Electrochim. Acta* 129: 55–61.
95. Zhao, P., Yang, J., Shang, Y., Wang, L., Fang, M., Wang, J., and He, X. 2015. Surface modification of polyolefin separators for lithium ion batteries to reduce thermal shrinkage without thickness increase. *J. Energy Chem.* 24(2): 138–44.
96. Rella, S., Giuri, A., Corcione, C. E., Acocella, M. R., Colella, S., Guerra, G., Listorti, A., Rizzo, A., and Malitesta, C. 2015. X-ray photoelectron spectroscopy of reduced graphene oxide prepared by a novel green method. *Vacuum* 119: 159–62.
97. Giuri, A., Rella, S., Malitesta, C., Colella, S., Listorti, A., Gigli, G., Rizzo, A., Cozzoli, P. D., Acocella, M. R., and Guerra, G. 2015. Synthesis of reduced graphite oxide by a novel green process based on UV light irradiation. *Sci. Adv. Mater.* 7(11): 2445–51.
98. Shi, C., Zhang, P., Chen, L., Yang, P., and Zhao, J. 2014. Effect of a thin ceramic-coating layer on thermal and electrochemical properties of polyethylene separator for lithium-ion batteries. *J. Power Sources* 270: 547–53.
99. Yang, P., Zhang, P., Shi, C., Chen, L., Dai, J., and Zhao, J. 2015. The functional separator coated with core-shell structured silica-poly-(methyl methacrylate) sub-microspheres for lithium-ion batteries. *J. Membr. Sci.* 474: 148–55.
100. Böhnstedt, W. 2004. A review of future directions in automotive battery separators. *J. Power Sources* 133(1): 59–66.
101. Chen, W., Shi, L., Wang, Z., Zhu, J., Yang, H., Mao, X., Chi, M., Sun, L., and Yuan, S. 2016. Porous cellulose diacetate-SiO_2 composite coating on polyethylene separator for high-performance lithium-ion battery. *Carbohydr. Polym.* 147: 517–24.
102. Chen, C., Xu, R., Chen, X., Xie, J., Zhang, F., Yang, Y., and Lei, C. 2016. Influence of cocrystallization behavior on structure and properties of HDPE/LLDPE microporous membrane. *J. Polym. Res.* 23(3): 1–9.
103. Hery, T., and Sundaresan, V. B. 2016. Ionic redox transistor from pore-spanning PPy(DBS) membranes. *Energy Environ. Sci.* 9: 2555–62.
104. Wang, X. Z., and Smela, E. 2009. Experimental studies of ion transport in PPy(DBS). *J. Phys. Chem. C* 113: 369–81.
105. Leontiadou, H., Mark, A. E., and Marrink, S. J. 2007. Ion transport across transmembrane pores. *Biophys. J.* 92(12): 4209–15.
106. West, B. J., Otero, T. F., Shapiro, B., and Smela, E. 2009. Chronoamperometric study of conformational relaxation in PPy(DBS). *J. Phys. Chem. B* 113: 1277–93.
107. Cao, L. Y., An, P., Xu, Z. W., and Huang, J. F. 2016. Performance evaluation of electrospun polyimide non-woven separators for high power lithium-ion batteries. *J. Electroanal. Chem.* 767: 34–9.
108. Zaccaria, M., Fabiani, D., Cannucciari, G., Gualandi, C., Focarete, M. L., Arbizzani, C., Giorgio, F. D., and Mastragostino, M. 2015. Effect of silica and tin oxide nanoparticles on properties of nanofibrous electrospun separators. *J. Electrochem. Soc.* 162(2): A915–20.
109. Chen, X., Chen, K., Wang, H., and Xue, D. 2015. A colloidal pseudocapacitor: Direct use of $Fe(NO_3)_3$ in electrode can lead to a high performance alkaline supercapacitor system. *J. Colloid Interface Sci.* 444: 49–57.

110. Chen, K., Xue, D., and Komarneni, S. 2015. Colloidal pseudocapacitor: nanoscale aggregation of Mn colloids from $MnCl_2$ under alkaline condition. *J. Power Sources* 279: 365–71.
111. EI-Kady, M. F., and Kaner, R. B. 2013. Scalable fabrication of high-power graphene micro-supercapacitors for flexible and on-chip energy storage. *Nat. Commun.* 4(2): 1475–5.
112. Liu, F., and Xue, D. 2013. An electrochemical route to quantitative oxidation of graphene frameworks with controllable C/O ratios and added pseudocapacitances. *Chem. Eur. J.* 19(32): 10716–22.
113. Zhang, J., Yue, L., Kong, Q., Liu, Z., Zhou, X., Zhang, C., Xu, Q., Zhang, B., Ding, G., Qin, B., Duan, Y., Wang, Q., Yao, J., Cui, G., and Chen, L. 2014. Sustainable, heat-resistant and flame-retardant cellulose-based composite separator for high-performance lithium ion battery. *Sci. Rep.* 4: 3935.
114. Huang, X. S. 2011. Separator technologies for lithium-ion batteries. *J. Solid State Electrochem.* 15(4): 649–62.
115. Chun, S. J., Choi, E. S., Lee, E. H., Kim J. H., Lee, S. Y., and Lee, S. Y. 2012. Performance evaluation of electrospun polyimide non-woven separators for high power lithium-ion batteries. *J. Mater. Chem.* 22(32): 16618–26.
116. Deimede, V., and Elmasides, C. 2015. Separators for lithium-ion batteries: a review on the production processes and recent developments. *Energy Technol.* 3(5): 453–68.
117. Tsao, C. H., Hsu, C. H., and Kuo, P. L. 2016. Ionic conducting and surface active binder of poly (ethylene oxide)-block-poly(acrylonitrile) for high power lithium-ion battery. *Electrochim. Acta* 196: 41–7.
118. Choi, J. A., Kim, S. H., and Kim, D. W. 2010. Enhancement of thermal stability and cycling performance in lithium-ion cells through the use of ceramic-coated separators. *J. Power Sources* 195: 6192–6.
119. Xie, Y., Zou, H. L., Xiang, H. F., Xia, R., Liang, D. D., Shi, P. C., Dai, S., and Wang, H. H. 2016. Enhancement on the wettability of lithium battery separator toward nonaqueous electrolytes. *J. Membrane Sci.* 503: 25–30.
120. Gong, Q., Yang, W., Jiang, L. and Ren, J. 2010. Porous Polyimide Membrane, Battery Separator, Battery, and Method. *US*20100028779A1.
121. Jabbour, L., Bongiovanni, R., Chaussy, D., Gerbaldi, C., and Beneventi, D. 2013. Cellulose-based Li-ion batteries: a review. *Cellulose* 20:1523–45.
122. Zolin, L., Destro, M., Chaussy, D., Penazzi, N., Gerbaldi, C., and Beneventi, D. 2015. Aqueous processing of paper separators by filtration dewatering: towards Li-ion paper batteries. *J. Mater. Chem. A* 3(28): 14894–901.
123. Zhang, J., Liu, Z., Kong, Q., Zhang, C., Pang, S., Yue, L., Wang, X., Yao, J., and Cui, G. 2013. Renewable and superior thermal-resistant cellulose-based composite nonwoven as lithium-ion battery separator. *ACS Appl. Mater. Interfaces* 5(1): 128–34.
124. Weng, B., Xu, F. H., Alcoutlabi, M., Mao, Y. B., and Lozano, K. 2015. Fibrous cellulose membrane mass produced via forcespinning® for lithium-ion battery separators. *Cellulose* 22(2): 1311–20.
125. Xu, Q., Kong, Q. S., Liu, Z. H., Zhang, J. J., Wang, X. J., Liu, R. Z., Yue, L. P., and Cui, G. L. 2014. Polydopamine-coated cellulose microfibrillated membrane as high performance lithium-ion battery separator. *RSC Adv.* 4 (16): 7845–50.
126. Missoum, K., Belgacem, M. N., and Bras, J. 2013. Nanofibrillated cellulose surface modification: a review. *Materials* 6(5): 1745–66.
127. Pan, R. J., Cheung, O., Wang, Z. H., Tammela, P., Huo, J. X., Lindh, J., Edström, K., Strømme, M., and Nyholm, L. 2016. Mesoporous Cladophora cellulose separators for lithium-ion batteries. *J. Power Sources* 321: 185–92.

128. Kuo, P. L., Wu, C. A., Lu, C. Y., Tsao, C. H., Hsu, C. H., and Hou, S. S. 2014. High performance of transferring lithium ion for polyacrylonitrile-interpenetrating crosslinked polyoxyethylene network as gel polymer electrolyte. *ACS Appl. Mater. Interfaces* 6(5): 3156–62.
129. Cao, J., Wang, L., He, X., Fang, M., Gao, J., Li, J., Deng, L., Chen, H., Tian, G., Wang, J., and Fan, S. J. 2013. In situ prepared nano-crystalline TiO_2-poly(methyl methacrylate) hybrid enhanced composite polymer electrolyte for Li-ion batteries. *J. Mater. Chem. A* 1(19): 5955–61.
130. Chen, K., Xue, D., and Komarneni, S. 2015. Beyond theoretical capacity in Cu-based integrated anode: insight into the structural evolution of CuO. *J. Power Sources* 275: 136–43.
131. Chen, K., and Xue, D. 2013. Room-temperature chemical transformation route to CuO nanowires toward high-performance electrode materials. *J. Phys. Chem. C* 117: 22576–83.
132. Hendricks, C., Williard. N., Mathew, S., and Pecht, M. 2015. A failure modes, mechanisms, and effects analysis (FMMEA) of lithium-ion batteries. *J. Power Sources* 297: 113–20.
133. Agubra, V. A., De la Garza, D., Gallegos, L., and Alcoutlabi, M. 2016. ForceSpinning of polyacrylonitrile for mass production of lithium-ion battery separators. *J. Appl. Polym. Sci.* 133: 42847.
134. Wu, W., Xiao, X., Huang, X., and Yan, S. 2014. A multiphysics model for the in situ stress analysis of the separator in a lithium-ion battery cell. *Comput. Mater. Sci.* 83: 127–36.
135. Venugopal, G., Moore, J., Howard, J., and Pendalwar, S. 1999. Characterization of microporous separators for lithium-ion batteries. *J. Power Sources* 77: 34–41.
136. Huang, X. 2011. Separator technologies for lithium-ion batteries. *J. Solid State Electrochem.* 15: 649–62.
137. Blanc, F., Leskes, M., and Grey, C. 2013. In situ solid-state NMR spectroscopy of electrochemical cells: batteries, supercapacitors, and fuel cells. *Acc. Chem. Res.* 46: 952–63.
138. He, M., Fic, K., Frackowiak, E., Novaka, P., and Berg, E. J. 2016. Ageing phenomena in high-voltage aqueous supercapacitors investigated by in situ gas analysis. *Energy Environ. Sci.* 9: 623–33.
139. Avasarala, B., Moore, R., and Haldar, P. 2010. Surface oxidation of carbon supports due to potential cycling under PEM fuel cell conditions. *Electrochim. Acta* 55: 4765–71.
140. Epstein, B. D., Dalle-Molle, E., and Mattson, J. S. 1971. Electrochemical investigations of surface functional groups on isotropic pyrolytic carbon. *Carbon* 9: 609–15.
141. Leyva-García, S., Morallón, E., Cazorla-Amorós, D., Béguin, F., and Lozano-Castelló, D. 2014. New insights on electrochemical hydrogen storage in nanoporous carbons by in situ Raman spectroscopy. *Carbon* 69: 401–8.
142. Wesełucha-Birczyńska, A., Babeł, K., and Jurewicz, K. 2012. Carbonaceous materials for hydrogen storage investigated by 2D Raman correlation spectroscopy. *Vib. Spectrosc.* 60: 206–11.
143. Jurewicz, K., Frackowiak, E., and Béguin, F. 2004. Towards the mechanism of electrochemical hydrogen storage in nanostructured carbon materials. *Appl. Phys. A: Mater. Sci. Process.* 78: 981–7.
144. Dubey, P. K., Sinha, A. S. K., Talapatra, S., Koratkar, N., Ajayan, P. M., and Srivastava, O. N. 2010. Hydrogen generation by water electrolysis using carbon nanotube anode. *Int. J. Hydrogen Energy* 35: 3945–50.

145. Yu, Y., Tu, Z., Zhang, H., Zhan, Z., and Pan, M. 2011. Comparison of degradation behaviors for open-ended and closed proton exchange membrane fuel cells during startup and shutdown cycles. *J. Power Sources* 196: 5077–83.
146. Fic, K., Meller, M., and Frackowiak, E. 2015. Interfacial redox phenomena for enhanced aqueous supercapacitors. *J. Electrochem. Soc.* 162: A5140–7.
147. Bratsch, S. G. 1989. Standard electrode potentials and temperature coefficients in water at 298.15 K. *J. Phys. Chem. Ref. Data* 18:1–21.
148. Huang, S., Hara, K., and Fukuoka, A. 2009. Green catalysis for selective CO oxidation in hydrogen for fuel cell. *Energy Environ. Sci.* 2: 1060–8.
149. Astarita, G. 1963. Absorption of carbon dioxide into alkaline solutions in packed towers. *Ind. Eng. Chem. Fundam.* 2: 294–7.
150. Kawamura, T., Kimura, A., Egashira, M., Okada, S., and Yamaki, J.-I. 2002. Thermal stability of alkyl carbonate mixed-solvent electrolytes for lithium ion cells. *J. Power Sources* 104: 260–4.
151. Wang, Q., Ping, P., Zhao, X., Chu, G., Sun, J., and Chen, C. 2012. Thermal runaway caused fire and explosion of lithium ion battery. *J. Power Sources* 208: 210–24.
152. Larush-Asraf, L., Biton, M., Teller, H., Zinigrad, E., and Aurbach, D. 2007. On the electrochemical and thermal behavior of lithium bis(oxalato)borate (LIBOB) solutions. *J. Power Sources* 174: 400–7.
153. Shigematsu, Y., Ue, M., and Yamaki, J.-I. 2009. Thermal behavior of charged graphite and LixCoO2 in electrolytes containing alkyl phosphate for lithium-ion cells. *J. Electrochem. Soc.* 156: A176–80.
154. Shigematsu, Y., Kinoshita, S.-I., and Ue, M. 2006. Thermal behavior of a C/$LiCoO_2$ cell, its components, and their combinations and the effects of electrolyte additives. *J. Electrochem. Soc.* 153: A2166–70.
155. Sinha, N. N., Burns, J. C., Sanderson, R. J., and Dahn, J. 2011. Comparative studies of hardware corrosion at high potentials in coin-type cells with non aqueous electrolytes. *J. Electrochem. Soc.* 158: A1400–3.
156. Kraemer, E., Schedlbauer, T., Hoffmann, B., Terborg, L., Nowak, S., Gores, H. J., Passerini, S., and Winter, M. 2013. Mechanism of anodic dissolution of the aluminum current collector in 1 M LiTFSI EC:DEC 3:7 in rechargeable lithium batteries. *J. Electrochem. Soc.* 160: A356–60.
157. Han, H.-B., Zhou, S.-S., Zhang, D.-J., Feng, S.-W., Li, L.-F., Liu, K., Feng, W.-F., Nie, J., Li, H., Huang, X. J., Armand, M., and Zhou, Z.-B. 2011. Lithium bis(fluorosulfonyl)imide (LiFSI) as conducting salt for nonaqueous liquid electrolytes for lithium-ion batteries: physicochemical and electrochemical properties. *J. Power Sources* 196: 3623–32.

5 Fabrication of Lithium-Ion Supercapacitors

Zhenyu Xing and Zhongwei Chen
University of Waterloo

CONTENTS

Lithium (Li)-ion supercapacitors have attracted researchers' attention with the combinations of high-energy density from Li-ion batteries and high power density from supercapacitors as shown in Figure 4.1 [1]. In this chapter, fabrication of Li-ion supercapacitors will be discussed in detail.

Generally, Li-ion supercapacitors are based on faradaic process and non-faradaic process to store Li-ions. The former involves Li-ion intercalation/deintercalation, while the latter one involves Li-ion sorption/desorption. Therefore, the electrode materials for the non-faradaic process are always like that of an electric double-layer capacitor, including porous graphene and activated carbon with high surface area. Further, the electrode materials for the faradaic process fall into the categories of the cathode materials and lithiated anode materials, such as layered oxides, spinel oxides, phosphates, fluorophosphates, silicates, hydroxides, transition metal oxides, polyanions, and graphite [1–7]. Proper selection of cathode material and anode material coupling with the carbonaceous material is very important to fabricate the Li-ion supercapacitor with high energy density, high power density, and superior cycling performance. Compared with the stable potential with a different state of charge for cathode materials and anode materials, the potential varies linearly with various charging stages for carbonaceous materials shown in Figure 5.1 [1]. Therefore, the following factors must be considered when fabricating Li-ion supercapacitors.

First, 0.5 V vs. Li or higher should be potential selection criteria for the faradaic electrode material, especially when Al foil is used as a current collector, to prevent Li–Al alloy from happening. Second, the overpotential should be low at high current density for the faradaic electrode; therefore, tuning the material size into nanoscale is a good option. Third, the faradaic electrode should give a super stable cycling performance up to 500,000 cycles to match the non-faradaic electrode. Fourth, the electrolyte should possess high ionic conductivity to maintain high power density.

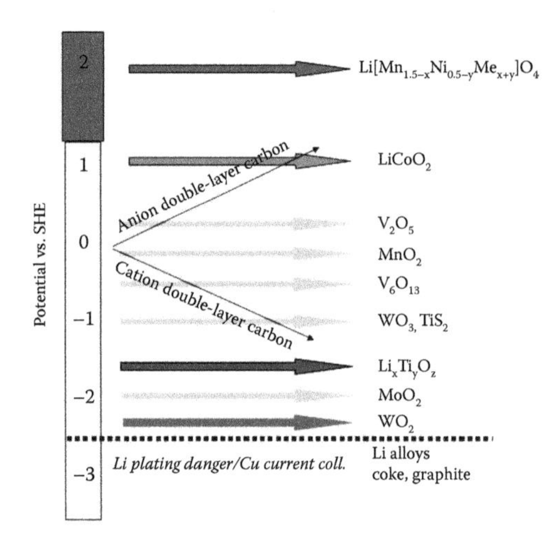

FIGURE 5.1 Schematic representation of three-electrode plot vs. SHE. The plot shows relative potentials of the positive and negative electrodes of an activated carbon electric double-layer capacitor as a function of charge versus general redox potential of various intercalation compounds discussed in the text [1]. Reprinted with permission from Ref. [1]. Copyright 2014 American Chemical Society.

Based on the aforementioned metrics to fabricate the Li-ion supercapacitor, we summarized a recent achievement of fabricating Li-ion supercapacitor with superior performance. Specifically, we will not focus on the material itself. Instead, the emphasis will be placed on the combination of the faradaic electrode and the non-faradaic electrode.

5.1 FARADAIC ANODE WITH CARBONACEOUS MATERIALS

Xia et al. reported β-FeOOH as a negative electrode and activated carbon as a positive electrode for Li-ion supercapacitor in 1 M $LiPF_6$ EC/DMC with a mole ratio of 1:2 [8]. Based on the capacity obtained from a three-electrode cell, the mass load of β-FeOOH and activated carbon is set to be 1–5. In the charging process, Li^+ cations are inserted into β-FeOOH, while PF_6^- anions are absorbed on the surface of activated carbon. Based on the total mass of two electrodes, the capacity is around 30 mAh/g and the energy density is around 45 Wh/kg. This hybrid capacitor showed a high rate performance, showing that the capacity retention can still be maintained around 80% even at 10 C current density. For the long cycling performance, 96% of original capacity is still maintained after 800 cycles.

As reported by Madhavi et al., V_2O_5 nanofibers prepared by electrospinning coupling with single-walled carbon nanotube network showed superior performance in 1 M $LiPF_6$ EC/DMC with a mole ratio of 1:1 [9]. Cyclic voltammograms (CVs) are performed in the potential range of 0–3 V; the specific capacitances are around 75, 51, 36, 24, 12, and 9 F/g based on a total mass of two electrodes when the various scanning rates sweep from 2, 5, 10, 20, 50 to 100 mV/s. Long cycling performance results are obtained by the galvanostatic charge–discharge process at a current density of 30 mA/g, and the initial capacitance is around 135 F/g. Though there is some capacitance fading at the beginning, the capacitance at 500th cycle can still be 109 F/g. As calculated, the peak energy density is 18 Wh/kg while the peak power density is 315 W/kg.

MnO_2 as a negative electrode and multi-walled carbon nanotubes as a positive electrode is studied in the Li-ion supercapacitor with 1.0M $LiClO_4$ in ethylene carbonate/diethyl carbonate (EC/DEC) by Yu et al [10]. The mass ratio between MnO_2 and carbon nanotubes is around 1:2.3. Under different current density, the peak capacitance is around 57 F/g. The energy density can reach 32.91 W h/kg under a current density of 10 mA/g. The capacity fades fast resulting in only a 70% capacity retention after 300 cycles. This may be due to the structure changes upon Li-ion intercalation/deintercalation, even the authors did not explain in the paper.

Kim et al. investigated the Li-ion supercapacitor composed of TiO_2@graphene as a negative electrode and activated carbon as a positive electrode in EC/DEC electrolyte containing 1M $LiPF_6$ [11]. As shown in Figure 5.2a, the CV curves of this hybrid capacitor are contributed by capacitive behavior from the activated carbon and redox behavior from TiO_2@graphene. The overall capacitance is around 89, 75, 63, and 53 F/g at a different scanning rate of 2.0, 5.0, 10.0, and 20.0 mV/s by CV in Figure 5.2b. The capacitance is further measured by galvanostatic charge technique in Figure 5.2c. At current density of 0.4, 0.8, 1.2, 1.6, 2.0, and 4.0 A/g, the capacitance is about 150, 90, 70, 55, 50, and 30 F/g (Figure 5.2d). The capacitance drops by 20% in the first 100 cycles but maintains a stable performance in the following cycles. High-energy density is around 42 Wh/kg when power density is 800 W/kg.

Deng et al. investigated $Li_2Ti_3O_7$ as a negative electrode and activated carbon as a positive electrode in the Li-ion hybrid capacitor [12]. The capacity is around 150 mAh/g only based on $Li_2Ti_3O_7$ mass when the hybrid capacitor is charged and discharged at a current density of 0.4 mA/cm^2. For the rate performance, the 75.9% capacitance can remain at a 10C rate when compared with a 1C rate. The overall cycling performance in 500 cycles is very stable with only a 5% capacitance loss.

Zheng et al. investigated $Li_4Ti_5O_{12}$ as an anode and activated carbon as a cathode in a three-electrode cell in which 1M $LiPF_6$ EC/DMC is used as the electrolyte [13]. To balance charge transfer, a mass ratio of 4:1 between activated carbon and $Li_4Ti_5O_{12}$ is adopted. As illustrated in Figure 5.3, the $Li_4Ti_5O_{12}$ possesses a two-phase reaction feature while activated carbon reveals a double-layer adsorption feature. The capacity is calculated to be 96 mAh/g when charging at 8C while discharging C/2. The $Li_4Ti_5O_{12}$/activated carbon hybrid capacitor shows an extremely stable cycling performance with 0% capacity being lost at an 8C charging rate. Even after cycling at a 10C rate following 4,500 cycles, 88% capacity retention is still maintained.

The $LiCrTiO_4$/activated hybrid Li-ion capacitor is also studied in 1M $LiPF_6$ (EC/DEC) with a mass ratio of 1:5 [14]. The electrochemical performance is evaluated in the potential range between 1 and 3V. Figure 5.4 summarized the cycling performance under charging rate from 0.4 to 2 A/g over 1,000 cycles. The highest capacity is 83 F/g under a current density of 0.4 A/g, whereas the lowest capacity is 16 F/g under 2 A/g. We find that the higher the current density is, the more stable cycling performance is. The improved capacity retention under higher current density may be due to less of a structure strain during Li-ion interaction/deintercalation.

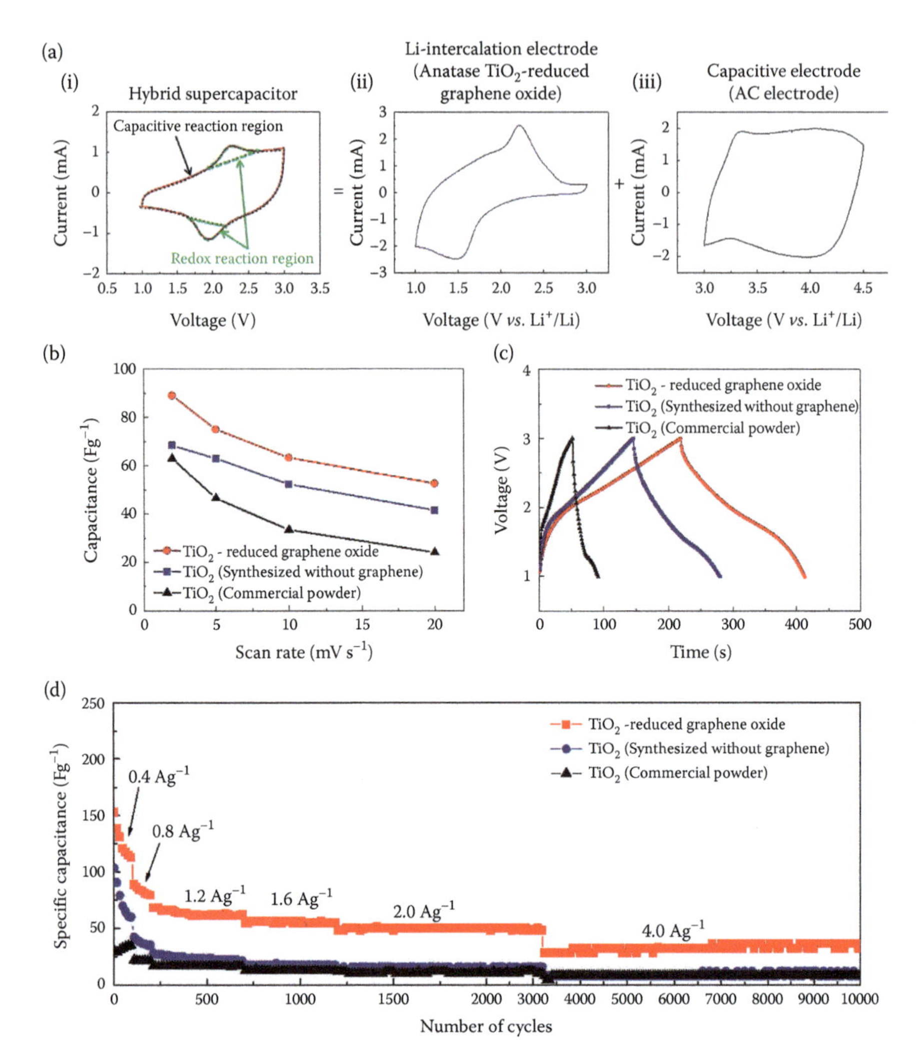

FIGURE 5.2 Electrochemical performance of TiO_2–(reduced graphene oxide) in the hybrid supercapacitor. (a) CV analysis of (i) the hybrid supercapacitor based on the TiO_2–(reduced graphene oxide) anode and AC cathode and of (ii, iii) TiO_2–(reduced graphene oxide) (ii) and AC electrodes (iii) in a Li half-cell. (b) The specific capacitance of TiO_2–(reduced graphene oxide) and TiO_2 nanoparticles in the hybrid supercapacitor. (c) Initial charge-discharge profiles of the hybrid supercapacitor. (d) Cyclability and rate capability of the hybrid supercapacitor at various current rates, from 0.4 to 4.0 A/g [11]. Reprinted with permission from Ref. [11]. Copyright 2013 Wiley-VCH Verbg GmbH & Co. KGaA.

TiP_2O_7 as an anode material and activated carbon as a cathode material are studied in the Li-ion capacitor by Aravindan et al. in 1 M $LiPF_6$ EC/DMC electrolyte [15]. Under different CV scans between 0 and 3 V, the capacitance based on total mass of anode and cathode is 21 F/g at 2 mV/s, 15 F/g at 5 mV/s, 10 F/g at 10 mV/s, 7 F/g at 20 mV/s, and 4 F/g at 50 mV/s. As indicated by the author, the

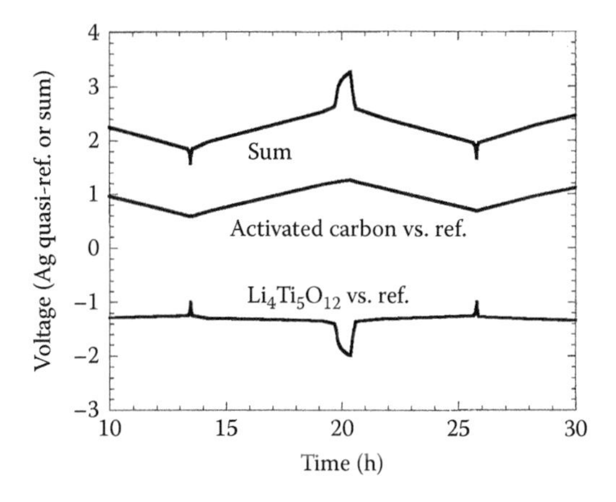

FIGURE 5.3 Three-electrode measurement of an asymmetric hybrid cell utilizing an activated carbon positive electrode and a $Li_4Ti_5O_{12}$ negative electrode in $LiPF_6$ EC/DMC electrolyte [13]. Reprinted with permission from Ref. [13]. Copyright 2001 The Electrochemical Society.

capacitance of a TiP_2O_7/activated carbon hybrid capacitor is much lower than that of an activated carbon/activated carbon double-layer capacitor. This is due to the insertion/extraction of Li-ion into/out of TiP_2O_7 is a bulk diffusion process, which greatly lowers the performance under high charging/discharging rate. The cycling performance is conducted at 31 mA/g, and the capacitance decreased from 38 to 30 F/g in the first 100 cycles, then keeps stable. From the Ragone plot, an energy density of 13 Wh/kg is obtained when the power density is 46 W/kg.

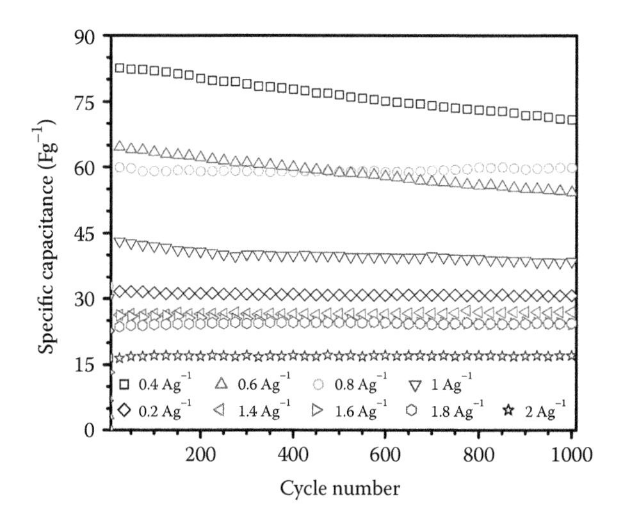

FIGURE 5.4 Plot of specific discharge capacitance versus cycle number of activated carbon/ $LiCrTiO_4$ Li-ion hybrid electrochemical capacitors (Li-HEC) with different current densities. The data points were collected after every 25 cycles [14]. Reprinted with permission from Ref. [14]. Copyright 2012, the Royal Society of Chemistry.

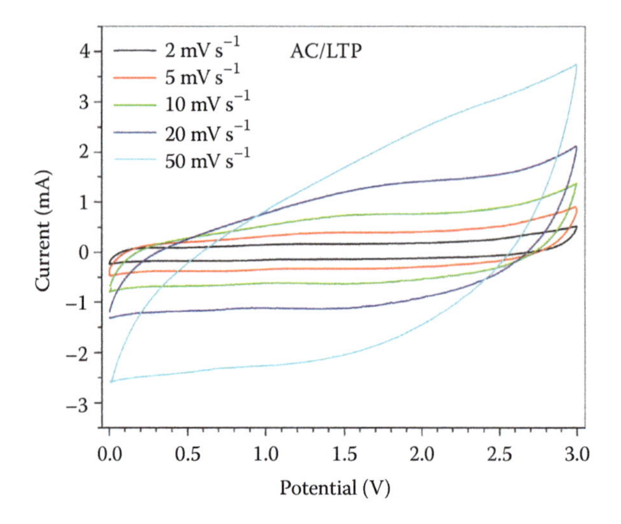

FIGURE 5.5 CV of the hybrid supercapacitor (HEC), activated carbon/carbon-coated nano-$LiTi_2(PO_4)_3$ (LTP) recorded between 0 and 3 V in a non-aqueous electrolyte (1 M $LiPF_6$ in EC: DMC 1: 1 V), at various sweep rates [16]. Reprinted with permission from Ref. [16]. Copyright 2012, the Royal Society of Chemistry.

NASICON-type $LiTi_2(PO_4)_3$ have also been paired up with activated carbon and investigated in the Li-ion supercapacitor. The electrolyte is 1 M $LiPF_6$ in EC/DMC [16]. As shown in the CV result, the rectangular shape can be maintained under low scanning rate while the shape is changed under high scanning rate. The specific capacitance is calculated to be 35, 30, 26, 21 and 14 F/g under different rate shown in Figure 5.5. The cycling performance is explored under a different current density from 120 to 60 mA/g. We can see that higher stability is maintained under higher current density, which is due to less of a structure strain under high current density with Li-ion only interacted into the near-surface region rather than the bulk phase.

Carbon-based capacitors usually store energy by electric double-layer adsorptions with a high surface area, while another type of carbon-based capacitor is utilizing Li-intercalated graphite as an anode and activated carbon as a cathode. The mechanism is shifted from physical adsorption to intercalation chemistry. As reported by Khomenko et al., this kind of hybrid capacitor is tested in the electrolyte of 1 M $LiPF_6$ in EC/DEC [17]. The galvanostatic charge–discharge curve is revealed by a three-electrode set up in Figure 5.6 in the potential range of 0–5 V. The curve of the hybrid capacitor is very similar to that of activated carbon rather than graphite, resulting from the higher current contribution of activated carbon. As we know, the intercalation of Li-ion into the graphite layer is very slow, causing the capacity to be dramatically dropped under high current density. However, the capacity of activated carbon will not be influenced by current density due to the surface adsorption mechanism. Thus, it is not difficult to understand why the energy density decreased while the power density increased. Further, the different mass ratio between lithiated graphite and activated carbon greatly affects the energy density and power density.

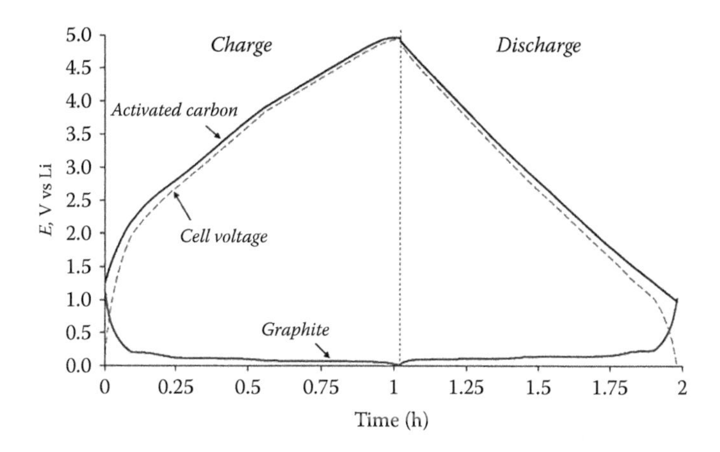

FIGURE 5.6 Galvanostatic charge–discharge at a current density C of the electrodes and cell (dotted line) for a hybrid capacitor utilizing activated carbon as positive electrode and graphite as a negative electrode. The cell was assembled with a mass ratio: m (activated carbon)/m (graphite) = 5/1 [17]. Reprinted with permission from Ref. [17]. Copyright 2008 Elsevier.

5.2 FARADAIC CATHODE WITH CARBONACEOUS MATERIALS

The layer oxide used as a cathode in Li-ion capacitor includes $LiNi_{1/3}Mn_{1/3}Fe_{1/3}O_2$, $LiNi_{1/3}Mn_{1/3}Co_{1/3}O_2$, and Li_2MoO_3. Since their structures are similar, here we only choose $LiNi_{1/3}Mn_{1/3}Fe_{1/3}O_2$ as an example to show how layer oxide is coupled with activated carbon as electrodes in the Li-ion capacitor. As reported by Lee et al., composite $LiNi_{1/3}Mn_{1/3}Fe_{1/3}O_2$ and polymer are used as a cathode in the Li-ion capacitor with 1 M $LiPF_6$ in EC/DMC [18]. $Li(Mn_{1/3}Ni_{1/3}Fe_{1/3})O_2$– polyaniline (PANI)/activated carbon shows the highest capacitance around 122 F/g compared with $Li(Mn_{1/3}Ni_{1/3}Fe_{1/3})O_2$–polypyrrole (PPy)/activated carbon, which is due to high electronic conductivity. The electrochemical performance of $LiNi_{1/3}Mn_{1/3}Fe_{1/3}O_2$ is further revealed by galvanostatic charge–discharge characterization. As shown in Figure 5.7a, there are two long slopes in the charge and discharge process, respectively. However, careful observation reveals that internal electric resistance will result in certain IR drop, interfacial charge transfer will result in a capacitive component and faradaic Li interaction will result in the long slope. The long cycling performance is quite stable for all three of $Li(Mn_{1/3}Ni_{1/3}Fe_{1/3})O_2$ for over 5,000 cycles (Figure 5.7b).

Typical spinel oxides as cathode material for the Li-ion supercapacitor are $LiMn_2O_4$ and $LiNi_{0.5}Mn_{1.5}O_4$, which possess higher potential feature than other oxides. Cericola et al. reported $LiMn_2O_4$/activated carbon compares remarkable performance in the continuous discharge and pulse discharge process, which is rarely studied by many researchers and deserves further investigation [19]. Ni-doped $LiNi_{0.5}Mn_{1.5}O_4$ is another promising cathode material, including two redox couples of Mn^{3+}/Mn^{4+} and Ni^{4+}/Ni^{2+} in two regions, in the Li-ion hybrid capacitor with activated carbon was studied by Xia et al. [20]. Based on the capacitance in the

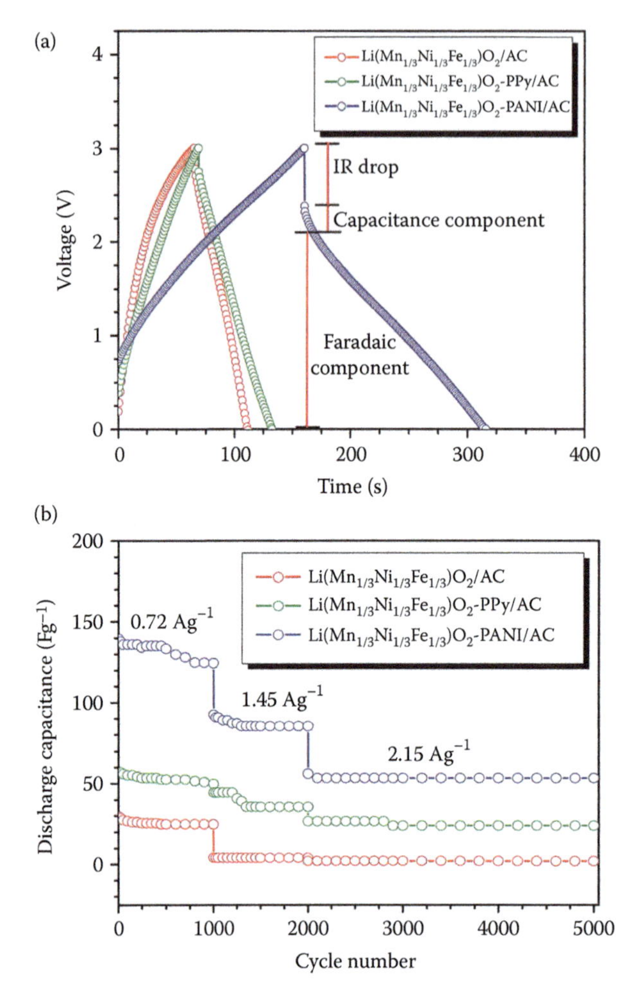

FIGURE 5.7 (a) Typical galvanostatic first charge–discharge curves of $Li(Mn_{1/3}Ni_{1/3}Fe_{1/3})O_2$/AC, $Li(Mn_{1/3}Ni_{1/3}Fe_{1/3})O_2$–PPy/AC, and $Li(Mn_{1/3}Ni_{1/3}Fe_{1/3})O_2$–PANI/AC cells measured at 0.72 A g1 between 0 and 3 V and (b) the cells' cycling profiles [18]. Reprinted with permission from Ref. [18]. Copyright 2008 Elsevier.

three-electrode cell, the mass ratio of $LiNi_{0.5}Mn_{1.5}O_4$ and activated carbon is set to be 1:3. The capacity of the hybrid capacitor is about 25 mAh/g, much higher than that of electric double-layer capacitor made of the same activated carbon. The rate capability is very attractive as the retention can be 90% when the current density is 10C. After 1,000 cycles, the total capacity only fades by 20%, suggesting a promising practical application.

Olivine phase $LiFePO_4$ and $LiCoPO_4$ are two typical phosphates cathodes in Li-ion capacitors. $LiFePO_4$ mixed with activated carbon as a cathode and mesocarbon microbeads as an anode in the Li-ion supercapacitor were studied by Wang et al. [21]. This hybrid capacitor delivers a capacity of 25 mAh/g when charging at a current density of 2C. However, the capacity is further lowered to 18.40 when the current density is increased to 10C. The capacity fades to 92.86% of original values after 100 cycles. In this work, the author primarily studied the influence of different activated carbons on the electrochemical performance of the Li-ion capacitor, indicating that electronic conductivity can greatly affect the polarization of the $LiFePO_4$, further affecting the capacity in the charge-discharge process. $LiCoPO_4$ is first introduced as a cathode material in the Li-ion supercapacitor by Renganathan et al. with carbon nanofoam as an anode in 1 M $LiClO_4$ EC/PC [22]. The highest energy density is about 11 Wh/kg at a power density of 192 W/kg. The total capacity dropped to 70% after 1,000 cycles, which called for further improvement.

An example of fluorophosphates used as cathode material in the Li-ion supercapacitor is Li_2CoPO_4F, which is coupled with activated carbon in 1 M $LiPF_6$ EC/DMC electrolyte [23]. The cyclic voltammetry is performed in the potential range between 0 and 3 V. Under various scanning rate from 2 to 15 mV/s, the curve remains a rectangular shape, indicating a capacitive feature. This is also why the capacity only dropped from 41 to 20 F/g when the scanning rate is almost 7 times larger. The remarkable rate capability is also confirmed by galvanostatic charge test. Under the current density of 1,100 mA/g, the cell shows barely any capacity fading, which is amazing.

Li_2MnSiO_4 and Li_2FeSiO_4, two representative silicates, are both investigated in the Li-ion supercapacitor with activated carbon by Karthikeyan et al. [24,25]. Li_2FeSiO_4 shows a bit higher capacity than Li_2MnSiO_4 at a current density of 1 mA/cm^2. After 1,000 cycles, the capacity retention for Li_2MnSiO is about 85% while that for Li_2FeSiO_4 is nearly 90%, which results from fewer structure changes for Li_2FeSiO_4 during the long cycling process.

This chapter mainly covers how the Li-ion capacitor is fabricated, especially introducing faradaic cathode materials and anode materials coupled with non-faradaic materials, which are activated carbons. However, the recent development on the non-faradaic side not only focuses on higher surfaces, more suitable porous structure, and higher conductivity of activated carbons, but also attempts to discover new carbonaceous material for higher electrical double-layer capacitance [26–29]. For example, graphene, carbon nanotubes, reduced graphene oxide, porous graphene, and their derived composite greatly expands the range of selection of non-faradaic materials for the Li-ion capacitor. The development and status of carbonaceous material go beyond the scope of this chapter and can be found widely in other review papers or books [30,31].

5.3 FABRICATION OF LI-ION SUPERCAPACITOR

Generally, the Li-ion supercapacitor fabrication includes three major steps: electrode preparation, cell assembly, and cell case sealing [32].

For electrode preparation, cathode material and anode material are grounded into a powder with a small particle size by a ball milling machine. Then, a mixer is used to mix the electrode material, binder, and conductive additive together at a specific ratio. The slurry coating is the most important step for electrode preparation. The

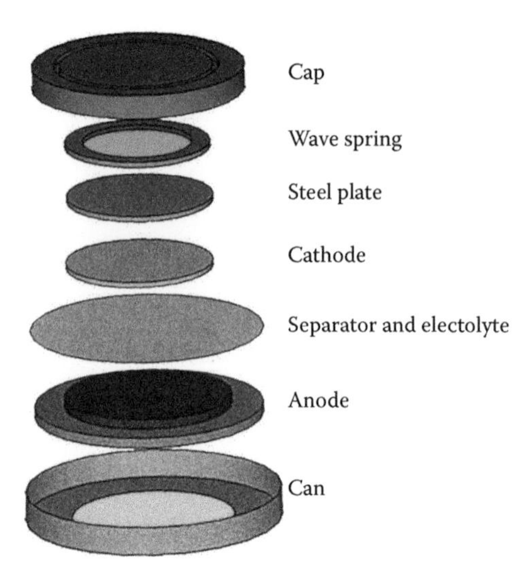

FIGURE 5.8 Assembly of coin cell for Li-ion supercapacitor [33].

doctor-blade method is widely used at lab scale while printing method is more used at industrial application. The reason why the printing method, such as spray printing, and inkjet printing, hold an advantage over the simple doctor-blade methods lies in the fact that printing can cover a very large surface. Another advantage of the printing method is not being critical of a specific substrate, common Al foil, Cu foil, or even soft polymer sheet, paper, and fabric, all of which can be used for printing method. After the ink is sprayed on the substrate, the solvent is evaporated after transfer the electrode into vacuum oven with elevated temperature. Here, it should be noted that the thickness of the electrode can be controlled by adjusting the amount of sprayed ink or rolling the electrode by the calendar [2].

After the electrodes are dried, they are cut into different shapes by cutting machine based on the requirement. In the lab, Li-ion supercapacitor is always assembled in the coin cell as shown in Figure 5.8. The anode material is always put on the can, and then the separator is put on the anode. Before the cathode is put on the top of the separator, a certain number of electrolytes are dropped in. Finally, the stainless steel spacer and the spring are put in with a cap covered. For the industrial application, the cells are always assembled in the form of pouch cell. Layers of anode, separator, cathode, and separator are stacked in a fixed order. The ultrasonic welding machine is then used to weld the tab and the current collector into a whole. Finally, the cell is dried again.

The last step to fabricate the Li-ion supercapacitor is by cell case sealing. This is very simple for coin cell because you simply press it under high pressure. However, sealing a pouch cell is a little more complex. The dried cell is put into a cup made by aluminum lamination sheet on which the cup is punched by a cup-forming machine. After one side is sealed, the electrolyte is filled in the glove box, followed by sealing the other side. Before the cells are sold to the market, a battery detector is used to check safety.

5.4 CHALLENGES AND FUTURE OUTLOOK

The sections above discussed full cell performance of the cathode/activated carbon or anode/activated carbon and summarize the fabrication procedure. Currently, the energy density and power density still cannot fulfill the requirement of the practical application. One primary reason behind this is because superior electrode materials are still lacking. Therefore, seeking for the cathode or anode material with high rate capability and stable cycling performance is very important. For the activated carbon, high energy density to be paired up with faradaic electrode is the direction researchers should follow. For practical application, profit is the core element every company cares about. The low-cost requirement will further make the task more challengeable.

However, we cannot be too pessimistic about the challenges because, at present, the intrinsic advantages of the Li-ion capacitor are still encouraging enough. Once the high energy density and high power density is obtained at the same time, the Li-ion supercapacitor will be a hot star in the energy storage field. Then, the portable electronics and electric vehicles will greatly benefit from the improvement in the Li-ion supercapacitor.

REFERENCES

1. Aravindan, V., Gnanaraj, J., Lee, Y.-S. & Madhavi, S. Insertion-type electrodes for nonaqueous Li-ion capacitors. *Chemical Reviews* **114**, 11619–11635 (2014).
2. Kashkooli, A. G., Farhad, S., Chabot, V., Yu, A. & Chen, Z. Effects of structural design on the performance of electrical double layer capacitors. *Applied Energy* **138**, 631–639 (2015).
3. Abureden, S. et al. Modified chalcogens with a tuned nano-architecture for high energy density and long life hybrid super capacitors. *Journal of Materials Chemistry A* **5**, 7523–7532 (2017).
4. Abureden, S. A., Hassan, F. M., Yu, A. & Chen, Z. Reconciled nanoarchitecture with overlapped 2D anatomy for high energy hybrid supercapacitors. *Energy Technology* **5**, 1919–1926 (2017).
5. Lee, S. W., Gallant, B. M., Byon, H. R., Hammond, P. T. & Shao-Horn, Y. Nanostructured carbon-based electrodes: bridging the gap between thin-film lithium-ion batteries and electrochemical capacitors. *Energy & Environmental Science* **4**, 1972–1985 (2011).
6. Wang, H., Zhu, C., Chao, D., Yan, Q. & Fan, H. J. Nonaqueous hybrid lithium-ion and sodium-ion capacitors. *Advanced Materials* **29**, 1702093–1702111 (2017).
7. Fan, X., Wang, X., Li, G., Yu, A. & Chen, Z. High-performance flexible electrode based on electrodeposition of polypyrrole/MnO_2 on carbon cloth for supercapacitors. *Journal of Power Sources* **326**, 357–364 (2016).
8. Cheng, L., Li, H.-Q. & Xia, Y.-Y. A hybrid nonaqueous electrochemical supercapacitor using nano-sized iron oxyhydroxide and activated carbon. *Journal of Solid State Electrochemistry* **10**, 405–410 (2006).
9. Aravindan, V. et al. Fabrication of high energy-density hybrid supercapacitors using electrospun V_2O_5 nanofibers with a self-supported carbon nanotube network. *ChemPlusChem* **77**, 570–575 (2012).
10. Wang, G.-X., Zhang, B.-L., Yu, Z.-L. & Qu, M.-Z. Manganese oxide/MWNTs composite electrodes for supercapacitors. *Solid State Ionics* **176**, 1169–1174 (2005).

11. Kim, H. et al. A novel high-energy hybrid supercapacitor with an anatase TiO_2–reduced graphene oxide anode and an activated carbon cathode. *Advanced Energy Materials* **3**, 1500–1506 (2013).
12. Chen, F. et al. Preparation and characterization of ramsdellite $Li_2Ti_3O_7$ as an anode material for asymmetric supercapacitors. *Electrochimica Acta* **51**, 61–65 (2005).
13. Amatucci, G. G., Badway, F., Du Pasquier, A. & Zheng, T. An asymmetric hybrid nonaqueous energy storage cell. *Journal of Electrochemical Society* **148**, A930–A939 (2001).
14. Aravindan, V., Chuiling, W. & Madhavi, S. High power lithium-ion hybrid electrochemical capacitors using spinel $LiCrTiO_4$ as insertion electrode. *Journal of Materials Chemistry* **22**, 16026–16031 (2012).
15. Aravindan, V. et al. Hybrid supercapacitor with nano-TiP_2O_7 as intercalation electrode. *Journal of Power Sources* **196**, 8850–8854 (2011).
16. Aravindan, V. et al. Carbon coated nano-$LiTi_2(PO_4)_3$ electrodes for non-aqueous hybrid supercapacitors. *Physical Chemistry Chemical Physics* **14**, 5808–5814 (2012).
17. Khomenko, V., Raymundo-Piñero, E. & Béguin, F. High-energy density graphite/AC capacitor in organic electrolyte. *Journal of Power Sources* **177**, 643–651 (2008).
18. Karthikeyan, K. et al. Unveiling organic–inorganic hybrids as a cathode material for high performance lithium-ion capacitors. *Journal of Materials Chemistry A* **1**, 707–714 (2013).
19. Cericola, D., Novák, P., Wokaun, A. & Kötz, R. Segmented bi-material electrodes of activated carbon and $LiMn_2O_4$ for electrochemical hybrid storage devices: effect of mass ratio and C-rate on current sharing. *Electrochimica Acta* **56**, 1288–1293 (2011).
20. Li, H., Cheng, L. & Xia, Y. A hybrid electrochemical supercapacitor based on a 5 V Li-ion battery cathode and active carbon. *Electrochemical and Solid-State Letters* **8**, A433–A436 (2005).
21 Ping, L., Zheng, J., Shi, Z., Qi, J. & Wang, C. Electrochemical performance of MCMB/ (AC+$LiFePO_4$) lithium-ion capacitors. *Chinese Science Bulletin* **58**, 689–695 (2013).
22. Vasanthi, R., Kalpana, D. & Renganathan, N. G. Olivine-type nanoparticle for hybrid supercapacitors. *Journal of Solid State Electrochemistry* **12**, 961–969 (2008).
23. Karthikeyan, K. et al. A high performance hybrid capacitor with Li_2CoPO_4 F cathode and activated carbon anode. *Nanoscale* **5**, 5958–5964 (2013).
24. Karthikeyan, K. et al. Electrochemical performance of carbon-coated lithium manganese silicate for asymmetric hybrid supercapacitors. *Journal of Power Sources* **195**, 3761–3764 (2010).
25. Karthikeyan, K. et al. A novel asymmetric hybrid supercapacitor based on Li_2FeSiO_4 and activated carbon electrodes. *Journal of Alloy and Compounds* **504**, 224–227 (2010).
26. Yu, A., Roes, I., Davies, A. & Chen, Z. Ultrathin, transparent, and flexible graphene films for supercapacitor application. *Applied Physics Letters* **96**, 253105 (2010).
27. Ahn, W. et al. Highly oriented graphene sponge electrode for ultra high energy density lithium ion hybrid capacitors. *ACS Applied Materials & Interfaces* **8**, 25297–25305 (2016).
28. Liu, Y., Li, G., Chen, Z. & Peng, X. CNT-threaded N-doped porous carbon film as binder-free electrode for high-capacity supercapacitor and Li–S battery. *Journal of Materials Chemistry A* **5**, 9775–9784 (2017).
29. Hassan, F. M. et al. Pyrrolic-structure enriched nitrogen doped graphene for highly efficient next generation supercapacitors. *Journal of Materials Chemistry A* **1**, 2904–2912 (2013).

30. Davies, A. et al. Graphene-based flexible supercapacitors: pulse-electropolymerization of polypyrrole on free-standing graphene films. *The Journal of Physical Chemistry C* **115**, 17612–17620 (2011).
31. Bokhari, S. W. et al. Nitrogen doping in the carbon matrix for Li-ion hybrid supercapacitors: state of the art, challenges and future prospective. *RSC Advances* **7**, 18926–18936 (2017).
32. Vangari, M., Pryor, T. & Jiang, L. Supercapacitors: review of materials and fabrication methods. *Journal of Energy Engineering* **139**, 72–79 (2012).
33. https://crystallography365.wordpress.com/2014/01/31/spinning-around-with-spinels-lithium-titanate/.

6 Theoretical Modeling of Lithium-Ion Supercapacitors

Dan Zhang and Delun Zhu
Shanghai University

CONTENTS

6.1 INTRODUCTION

Lithium-ion capacitors (LiCs) belong to typical asymmetric electrode designs, which are most frequently associated to the combination of a capacitive electric double-layer (EDL) electrode and a battery-type faradaic or pseudocapacitive electrode. When the battery-type electrode is a Li-intercalating phase, these systems are then referred to as LiCs [1]. So far, the intense modeling activity is witnessed by quite extensive literature on lithium-ion batteries (LiBs) and electric double-layer capacitors (EDLCs). For example, models for LiBs of different technologies can be found in [2–11], and supercapacitor (SC) models are proposed in [12–20]. Modeling of LiCs is a full challenge that has not yet been receiving a great interest so far, but will be of high expectation due to two main commercial drives: the great push for both intermittent energy storage devices by the renewable energy systems and hybrid electric vehicles by the automotive industry, and the urgent desire for new simulation-based development methods dealing with the complexity of modern power electronic systems [21–25].

Providing high energy density, high power density, high efficiency, and long cycle life, a LiC is a hybrid of an EDLC and a LiB [26,27]. Many of the techniques developed for modeling LiBs and EDLCs are applicable to the more modern chemistries involved in LiCs. However, more efforts need to be made in order to drive down the cost of LiCs, and to give a deep insight into the dynamic voltage behavior and energy efficiency particularly at highly dynamic current profiles and relatively low or high temperatures [21,28].

In this chapter, we consider both circuit-based cell models and physics-based cell models and focus ourselves on the most common LiC chemistries and modeling techniques, which means LiCs use voltage, current, concentration, temperature, and resistance-capacitance (*RC*) as the physical quantities of interest, and can be written as the equivalent impedance equations.

This chapter is organized as follows: Section 6.2 summaries approaches to the LiC modeling as they apply to various SC modeling tasks. Section 6.3 focuses on equivalent circuit models in particular. The chapter is finished off by the references.

6.2 MODELING LITHIUM-ION SUPERCAPACITORS

In order to use LiC in renewable energy systems, or electric vehicles, dynamic characteristics of LiC have to be identified first of all, therefore appropriate dynamic models are necessary for modeling and simulations of systems containing LiC. Several approaches to SC dynamic modeling are applicable to LiCs. The first group of models consists of equivalent circuit models [13,29–33]. The second class of SC models is physical-based models [14,34]. A special class of models combine elements of both chemistry and circuit-based modelling techniques [35,36]. Another uncommon SC model is black-box based models such as artificial neural networks [17]. The first is the most common technique we encountered for modeling LiCs in renewable energy systems and automotive applications.

Equivalent circuit models abstract away the electrochemical nature of the electrochemical cells and represent it solely as electrical components [37]. These equivalent circuit models are usually proposed with combinations of series and parallel *RC* circuits, which can lead to accurate voltage–current behavior on the terminals of a SC. Sometimes these components also contain inductances and nonlinear components like diodes that strive to better approximate the electrochemical nature of the cells [22,23]. Model parameters including resistances and capacitances are usually determined with the help of electrochemical impedance spectroscopy (EIS) measurements of currents and voltages during charging and discharging processes. The EIS approach largely based on similar methods used to analyze circuits in electrical engineering practice [38–41] deals with the variation of total impedance of electrochemical cells in the complex plane (Nyquist plots) by an equivalent circuit consisting of resistors and capacitors that pass current with the same amplitude and phase angle that the real cell does under a given excitation. This method circumvents the need for separate measurements without the electroactive species and also eliminates the need to assume that the electroactive species has no effect on the non-faradaic impedance [42].

The EIS measurements varying with frequency are capable of high precision and are frequently used for the extraction of the frequency dependencies of those circuit components involved in the equivalent circuit. The EIS technique can not only monitor changes in the electrochemical cell under different usage or storage conditions but can also provide detailed kinetic information such as the heterogeneous charge-transfer parameters for studies of the electrode structure. For a better understanding of the EIS methodology, in the following section, we depict the structure of a novel hybrid electrode, propose its corresponding equivalent circuit and summary necessary theory that is applied to interpret the equivalent resistance and capacitance values in terms of interfacial phenomena of the hybrid electrode.

Figure 6.1a is the illustration of possible hybrid devices consisting in turn of battery and capacitor components [1]. Figure 6.1b illustrates a possible hybrid electrode design of double hybrid devices (as shown in Figure 6.1a center), underlying both double-layer and pseudocapacitive mechanisms in a single electrode. The double

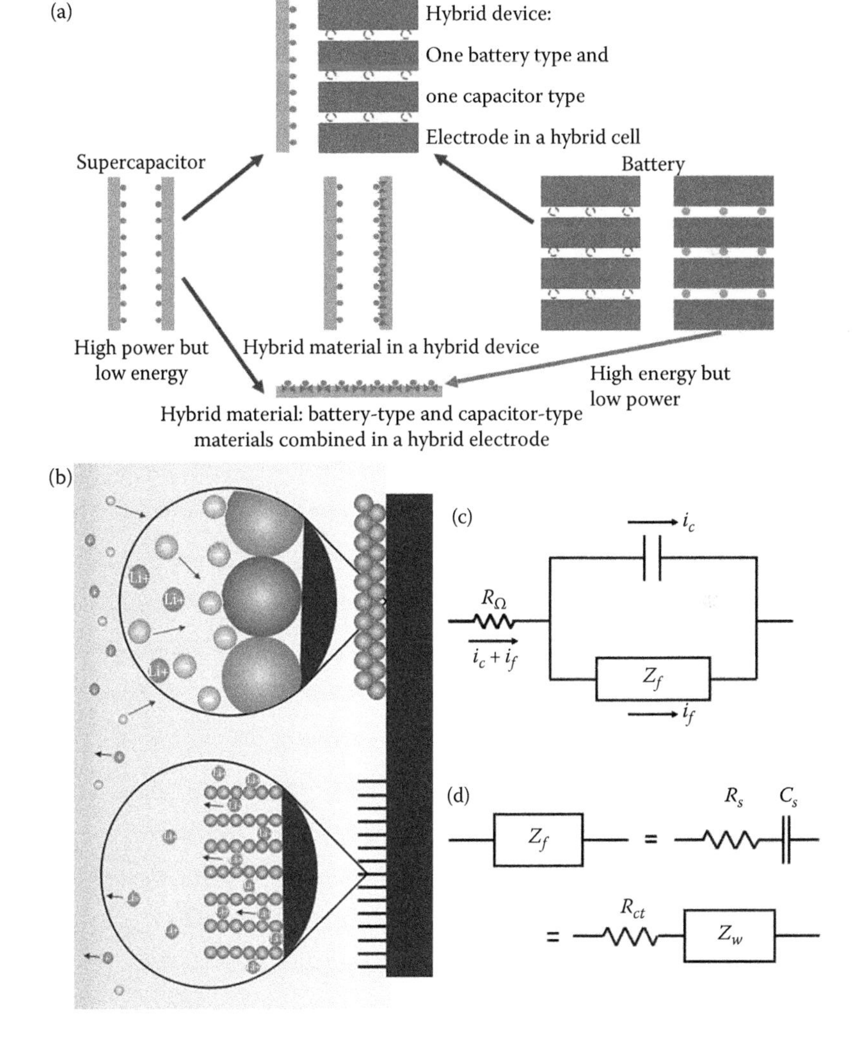

FIGURE 6.1 (a) Schematic representations of different possible hybridization approaches between a supercapacitor and battery electrodes and materials [1]. (b) A hybrid electrode underlying both double-layer and pseudocapacitive mechanisms. (c) Equivalent circuit of the hybrid electrode. (d) Subdivision of Z_f into R_s and C_s, or into R_{ct} and Z_w.

layer mechanism shown here arises from adsorption of negative ions from the electrolyte on the positively charged electrode. The pseudocapacitive mechanism results from deintercalation of Li-ions that are inserted in advance into the host electrode material [27]. A circuit proposed to model this hybrid electrode, called the *Randle equivalent circuit*, is shown in Figure 6.1c. The parallel elements represent the total current through the hybrid electrode, i.e., the sum of distinct contributions from the faradaic process, i_f, and double-layer charging, i_c. The double-layer capacitance is nearly a pure capacitance, of which the equivalent circuit element is C_d. The faradaic process depends on the current frequency, ω, which has to be considered as a general impedance, Z_f. The series element R_Ω of the equivalent circuit represents the Ohmic resistance of the hybrid electrode that all of the current must pass through, including the contact resistance of the electrode and the resistance of the electrolyte. Two equivalences to the general impedance, Z_f, are shown in Figure 6.1d. The faradaic impedance, Z_f, can be regarded as a series combination of the polarization resistance, R_s, and the pseudocapacity, C_s, or as a series combination of a pure charge-transfer resistance, R_{ct}, and another general impedance, Z_w, the Warburg impedance, which represents a kind of resistance to mass transfer.

Now consider that the measured EIS total impedance, Z, of the hybrid electrode shown in Figure 6.1c is expressed as the series combination of R_B and C_B, which provide the real component, Z_{Re}, and the imaginary components, Z_{Im}:

$$Z = R_B - j/(\omega C_B) = Z_{\text{Re}} - jZ_{\text{Im}}. \tag{6.1}$$

Substituting the circuit components shown in Figure 6.1c and d into Eq. 6.1 to yield

$$Z_{\text{Re}} = R_\Omega + \frac{R_s}{A^2 + B^2}, \quad Z_{\text{Im}} = \frac{B^2/\omega C_d + A/\omega C_s}{A^2 + B^2} \quad \text{and} \tag{6.2}$$

$$A = \left(C_d/C_s\right) + 1, \quad B = \omega R_s C_d \tag{6.3}$$

Consider the behavior of the faradaic impedance as a sinusoidal current, i_f, is forced through it. The total voltage drop on the impedance is

$$E = iR_s + \frac{q}{C_s}, \quad i_f = I_{f0} \sin \omega t, \quad \frac{di_f}{dt} = I_{f0}\omega \cos \omega t. \tag{6.4}$$

Hence, the differential of E to time, t, can be written as the following electric equation

$$\frac{dE}{dt} = \left(R_s I_{f0}\omega\right)\cos \omega t + \left(\frac{I_{f0}}{C_s}\right)\sin \omega t. \tag{6.5}$$

Consider a standard faradaic process on the cell electrode, $O + ne \rightleftharpoons R$, with a standard rate constant, k^0, the transfer coefficient, α, and the exchange current, i_0, we can write the voltage drop as a function of the current stimulus, i_f, and the concentrations, C_O and C_R

$$E = E_{(i_f, C_O, C_R)} \tag{6.6}$$

Hence, the differential of E to time, t, can be written as the chemical equation

$$\frac{dE}{dt} = \left(\frac{\partial E}{\partial i_f}\right)\frac{di_f}{dt} + \left(\frac{\partial E}{\partial C_O}\right)\frac{dC_O}{dt} + \left(\frac{\partial E}{\partial C_R}\right)\frac{dC_R}{dt} \tag{6.7}$$

Consider R_{ct} is the charge-transfer resistance, substituting it into Eq. 6.7 to yield

$$\frac{dE}{dt} = R_{ct}\frac{di_f}{dt} + \beta_O\frac{dC_O}{dt} + \beta_R\frac{dC_R}{dt} \tag{6.8}$$

Consider the linearized solution of the current-overpotential equation, it yields

$$E - E_{eq} = \frac{RT}{nF}\left(\frac{i_f}{i_0} + \frac{C_O}{C_O^*} - \frac{C_R}{C_R^*}\right) \tag{6.9}$$

where E_{eq} is the equilibrium potential of the electrode. Equating the coefficients of those terms of Eq. 6.8 with those of the derivatives of Eq. 6.9 to yield

$$\begin{aligned} R_{ct} &= \frac{\partial E}{\partial i_f} = \frac{RT}{nFi_0}, \\ \beta_O &= \frac{\partial E}{\partial C_O} = \frac{RT}{nFC_O^*}, \quad \text{and} \\ \beta_R &= \frac{\partial E}{\partial C_R} = -\frac{RT}{nFC_R^*}, \end{aligned} \tag{6.10}$$

where i_0 is the exchange current and F is Faraday's constant. Consider semi-infinite linear diffusion with initial conditions $C_{O(x,0)} = C_O^*$ and $C_{R(x,0)} = C_R^*$, the Taylor expansion of the reaction-diffusion equations of the concentrations to yield the derivatives of C_O, C_R, near the surface,

$$\begin{aligned} \frac{dC_O}{dt} &= \frac{I_{f0}}{nFA}\left(\frac{\omega}{2D_O}\right)^{\frac{1}{2}}(\sin\omega t + \cos\omega t) \\ \frac{dC_R}{dt} &= -\frac{I_{f0}}{nFS}\left(\frac{\omega}{2D_R}\right)^{\frac{1}{2}}(\sin\omega t + \cos\omega t), \end{aligned} \tag{6.11}$$

where S is the electrode surface area. Substituting Eqs. 6.4, 6.10, and 6.11 into Eq. 6.8 to yield

$$\frac{dE}{dt} = \left(R_{ct} + \frac{\sigma}{\omega^{1/2}}\right)I_{f0}\omega\cos\omega t + I_{f0}\sigma\omega^{1/2}\sin\omega t \quad \text{and} \tag{6.12}$$

$$\sigma = \frac{1}{\sqrt{2}nFS}\left(\frac{\beta_O}{\sqrt{D_O}} - \frac{\beta_R}{\sqrt{D_R}}\right). \tag{6.13}$$

Comparing Eq. 6.12 with Eq. 6.5 to yield

$$R_s = R_{ct} + \sigma\omega^{-1/2}, \quad C_s = 1/\left(\sigma\omega^{1/2}\right) \quad \text{and} \tag{6.14}$$

$$Z_f = R_{ct} + \sigma\omega^{-1/2} - j\left(\sigma\omega^{-1/2}\right), \quad Z_w = \sigma\omega^{-1/2} - j\left(\sigma\omega^{-1/2}\right). \tag{6.15}$$

Substituting Eq. 6.14 into Eqs. 6.3 and 6.2 to yield

$$\begin{aligned} Z_{\mathrm{Re}} &= R_\Omega + \frac{R_{ct} + \sigma\omega^{-1/2}}{\left(\sigma\omega^{1/2}C_d + 1\right)^2 + \omega^2 C_d^{\,2}\left(R_{ct} + \sigma\omega^{-1/2}\right)^2} \\ Z_{\mathrm{Im}} &= \frac{\omega C_d\left(R_{ct} + \sigma\omega^{-1/2}\right)^2 + \sigma\omega^{-1/2}\left(\sigma\omega^{1/2}C_d + 1\right)}{\left(\sigma\omega^{1/2}C_d + 1\right)^2 + \omega^2 C_d^{\,2}\left(R_{ct} + \sigma\omega^{-1/2}\right)^2}. \end{aligned} \tag{6.16}$$

As frequency-response analysis methods, Nyquist plots are widely used to extract chemical information of the electrochemical cell by plotting Z_{Re} vs. Z_{Im} for different ω. An illustration for the hybrid electrode of Figure 6.1c is shown in Figure 6.2.

It is worth noting that two regions of mass-transfer and kinetic control are recognized at low and high frequencies, respectively. For simplicity, let's consider the limiting behavior of the electrochemical cell at high and low ω. At the low-frequency limit, Z_{Re} and Z_{Im} of Eq. 6.16 approach their limiting forms as the frequency, ω, approaches zero.

$$\begin{aligned} Z_{\mathrm{Re}} &= R_\Omega + R_{ct} + \sigma\omega^{-1/2} \\ Z_{\mathrm{Im}} &= 2\sigma^2 C_d - \sigma\omega^{-1/2}. \end{aligned} \tag{6.17}$$

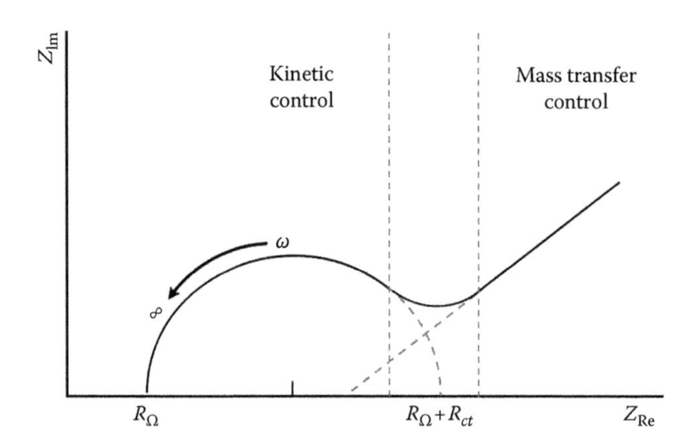

FIGURE 6.2 An illustration of Nyquist plots for the electrochemical system of Figure 6.1b underlying both mass-transfer and kinetic control mechanisms.

Hence,

$$Z_{\text{Im}} = Z_{\text{Re}} - \left(R_{\Omega} + R_{ct} - 2\sigma^2 C_d\right) \tag{6.18}$$

It is clear that the plot of Eq. 6.18 should be linear and have a unit slope, corresponding to the mass-transfer control region as shown in Figure 6.2. The extrapolated line intersects the real axis at $R_{\Omega} + R_{ct} - 2\sigma^2 C_d$. The frequency dependence in this region comes only from Warburg impedance terms; thus the linear correlation is characteristic of a diffusion-controlled electrode process. For example, for a LiC, it is the solid diffusion of Lithium-ion intercalation in the anode material that controls the electrode process.

With increasing frequency, we can expect a departure from the linear region and at very high frequencies, the Warburg impedance becomes unimportant; thus, Z_{Re} and Z_{Im} of Eq. 6.16 can be simplified into

$$Z_{\text{Re}} = R_{\Omega} + \frac{R_{ct}}{1+\omega^2 C_d^2 R_{ct}^2} \quad \text{and}$$
$$Z_{\text{Im}} = \frac{\omega C_d R_{ct}^2}{1+\omega^2 C_d^2 R_{ct}^2}. \tag{6.19}$$

Hence,

$$\left(Z_{\text{Re}} - \left(R_{\Omega} + \frac{R_{ct}}{2}\right)\right)^2 + Z_{\text{Im}}^2 = \left(\frac{R_{ct}}{2}\right)^2. \tag{6.20}$$

Here, at high frequencies, the charge-transfer resistance, R_{ct} and the double-layer capacitance, C_d, become more important and the equivalent circuit converges to the semicircle, with the center at $Z_{\text{Im}} = 0$ and $Z_{\text{Re}} = R_{\Omega} + R_{ct}/2$ and a radius of $R_{ct}/2$, as shown in the kinetic control region of Figure 6.2.

In fact, both regions may not be well defined for any real system. For a kinetically sluggish chemical system, a large R_{ct} will appear and mass transfer is a significant factor only in a very limited-frequency region. On the contrary, for a kinetically facile system, a small R_{ct} compared to R_{Ω} and Z_w might emerge and mass transfer always plays a role over nearly the whole available frequency range; thus the semicircular region is not well defined.

6.3 EQUIVALENT CIRCUIT MODELS

In order to build an appropriate equivalent circuit model of the LiC system, we should have a clear grasp of cell configurations prepared for the EIS measurement. LiCs can be analyzed by either a two-terminal or a three-terminal configuration, as shown in Figure 6.3 [43]. The two-terminal test is normally used to measure a whole LiC or cell, while a reference electrode is introduced to study the chemistry at each individual electrode during a three-terminal configuration. For convenience, the two electrodes of LiCs are named as the SC-EDL electrode and LiB electrode,

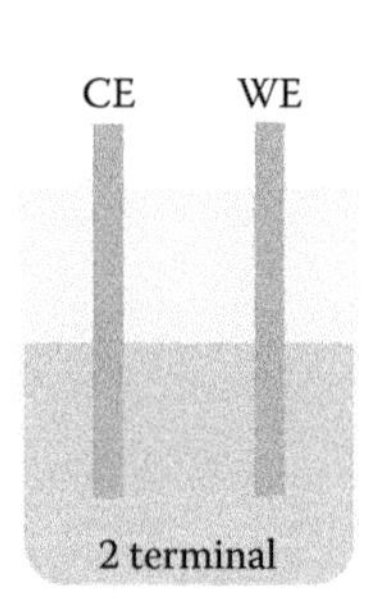

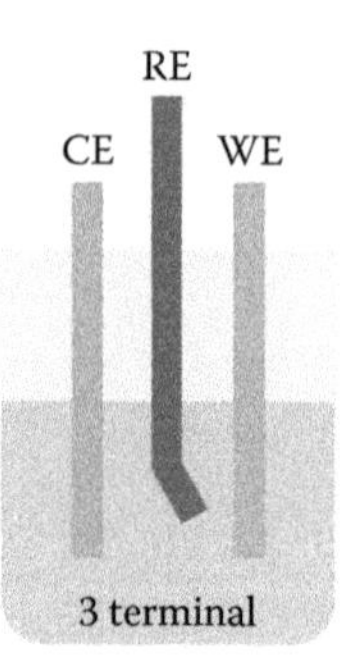

FIGURE 6.3 Illustration of typical cell configurations for a whole LiC cell (a) or an individual electrode (b).

respectively. In the next sections, we will depict equivalent circuit models in terms of an individual electrode and a whole LiC, respectively.

6.3.1 Models for Supercapacitor-Electric Double-Layer Electrode

One of the two LiC electrodes normally has similar structure with the EDL electrode of typical SCs. Usually, SCs are modeled using simple *RC* circuits, but these simple models cannot accurately describe the voltage behavior and the energy efficiency of these devices during highly dynamic current profiles [44]. Numerical simulations show the wide-band frequency response of SC can be well represented by a five-cascaded *RC* circuits, as shown in Figure 6.4 [45,46].

Although N-interleaved *RC* circuits would lead to satisfying results, it is nearly impossible to determine more than five-cascaded parameters in an efficient way because these N-interleaved parameters have a strong influence on each other [16]. A more advanced equivalent circuit model is proposed to model the electric behavior of SCs in highly dynamic applications by Buller, etc., which consists of an inductor to avoid fitting errors in the intermediate frequency, a series resistor and a complex impedance (Z_p) responsible for diffusion in porous electrode, as shown in Figure 6.5a [16].

The above is a mathematically simple model because only four parameters need to be determined, i.e., L, R_i, C, and t. Figure 6.5 gives an approximation of Z_p through

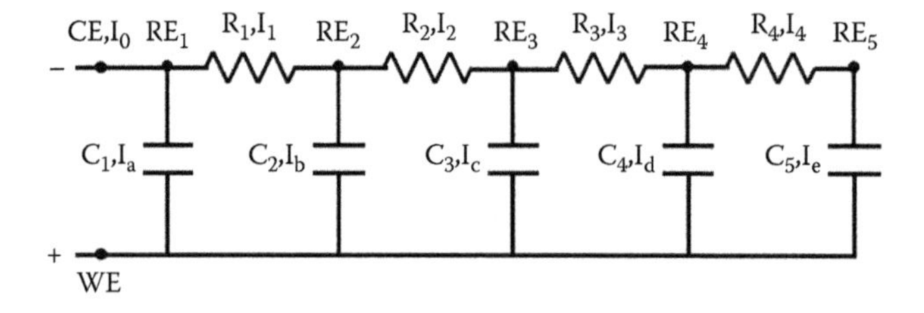

FIGURE 6.4 An N-cascaded equivalent *RC* circuit proposed for a porous supercapacitor electrode [45,46].

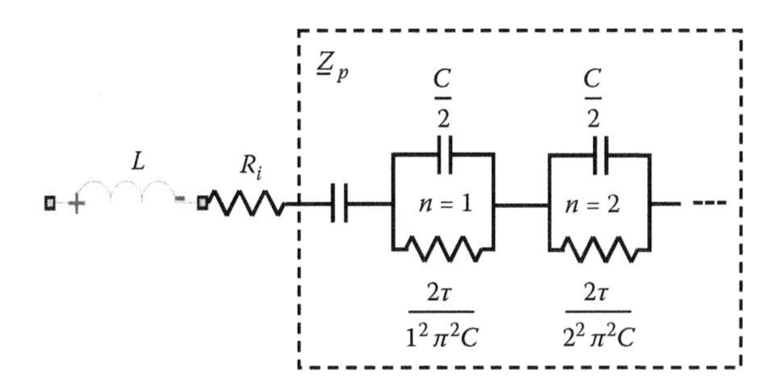

FIGURE 6.5 An equivalent circuit for a typical porous electrode SC.

N-*RC* circuits, which is a function of two independent parameters (C, τ). The expression of Z_p is listed in the following equation:

$$Z_p = \frac{\tau \cdot \coth\left(\sqrt{j\omega T}\right)}{C \cdot \sqrt{j\omega T}}. \tag{6.21}$$

However, the model mentioned above isn't able to reproduce the self-discharge phenomenon of SCs and is not accurate at very low frequencies too [47], which leads to the request for more sophisticated models. Considering the recombination phenomena after fast charge or discharge, and self-discharge phenomena, Musolino proposed an improved two-branch model based on Tironi's work, as shown in Figure 6.6 [47,48].

In this model, besides the voltage-dependent impedance Z_p, two additional parallel sub-branches are sufficient to achieve accurate simulation results, including a series *RC* branch, which represents the recombination phenomena after fast charge

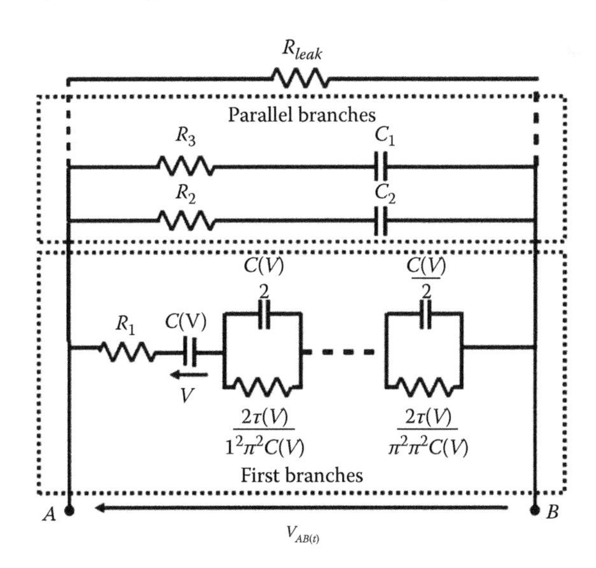

FIGURE 6.6 An improved two-branch circuit for a porous electrode SC [47,48].

or discharge, and a leakage resistance that takes into account the self-discharge phenomena. The inductive behavior is not interesting for a SC application; therefore, the inductor component is neglected from the improved two-branch model. Good agreement of the model prediction with a set of experiments on different manufactured cells verifies that the improved model can simulate both the dynamic response and the very slow phenomena that occur inside SCs; in other words, it can represent the behavior of SCs over the entire frequency range, from dc to tens of kilohertz.

6.3.2 Models for Lithium-Ion Battery Electrode

Electrochemical kinetic characteristics for battery materials are represented by several common steps, including ionic and electronic conduction through layer of active material, reaction on the particle interface, diffusion inside particles and phase-transfer reaction if in case several phases are present. Each step of Li-ion insertion/extraction into/out of anode and cathode of LiBs can be depicted with a electric component. Two synchronous steps can be equivalently expressed by two electric components with parallel connection, while two successive steps by two electric components with series connection.

Aurbach and coworkers studied the intercalation mechanism of Li-ion into thin graphite-coated electrodes with the aim of better distinguishing between its different relaxation steps [49]. They conducted simultaneously CV and EIS characterization of the intercalation of lithium into graphite and found that Li-ion intercalation into graphite involves several processes which occur in series: diffusion of Li ions in solutions, charge transfer, migration of Li-ions through the surface films (solid–electrolyte interface (SEI)), which cover the graphite particles, solid-state diffusion of Li-ions into the graphite, and finally, accumulation consumption of Li-ion into graphite, which accompanies phase transition between intercalation stages. Figure 6.7 describes the equivalent circuit analog Aurbach

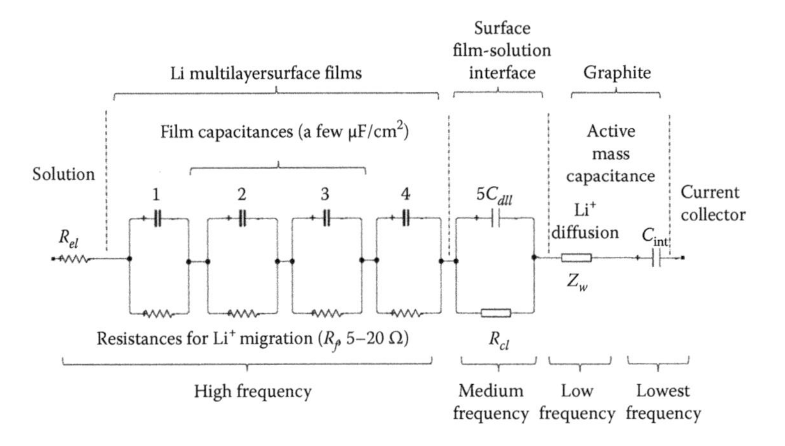

FIGURE 6.7 Equivalent circuit used for analysis of impedance spectra of the lithium-ion insertion/desertion in the intercalation electrode [49].

proposed, which seems to well present the lithium–graphite intercalation, based on a combination of a Voigt-type analog and the generalized Frumkin and Melik-Gaykazyan impedance.

The abovementioned equivalent circuit model is applicable to investigate the kinetics of Li-ion intercalation into thin carbon layers with a well-defined structure. To study the kinetics involved in Li-ion intercalation into mesoporous electrode materials, Barsoukov and coworkers proposed a kinetic model, taking into account both diffusion in a single particle and new phase formation in crystals of intercalation materials during low frequency impedance measurements, as shown in Figure 6.8a [50]. Figure 6.8b shows the equivalent circuit describing small-voltage behavior of battery insertion electrodes, formulated in normalized form in terms of electric elements as shown in Eq. 6.22.

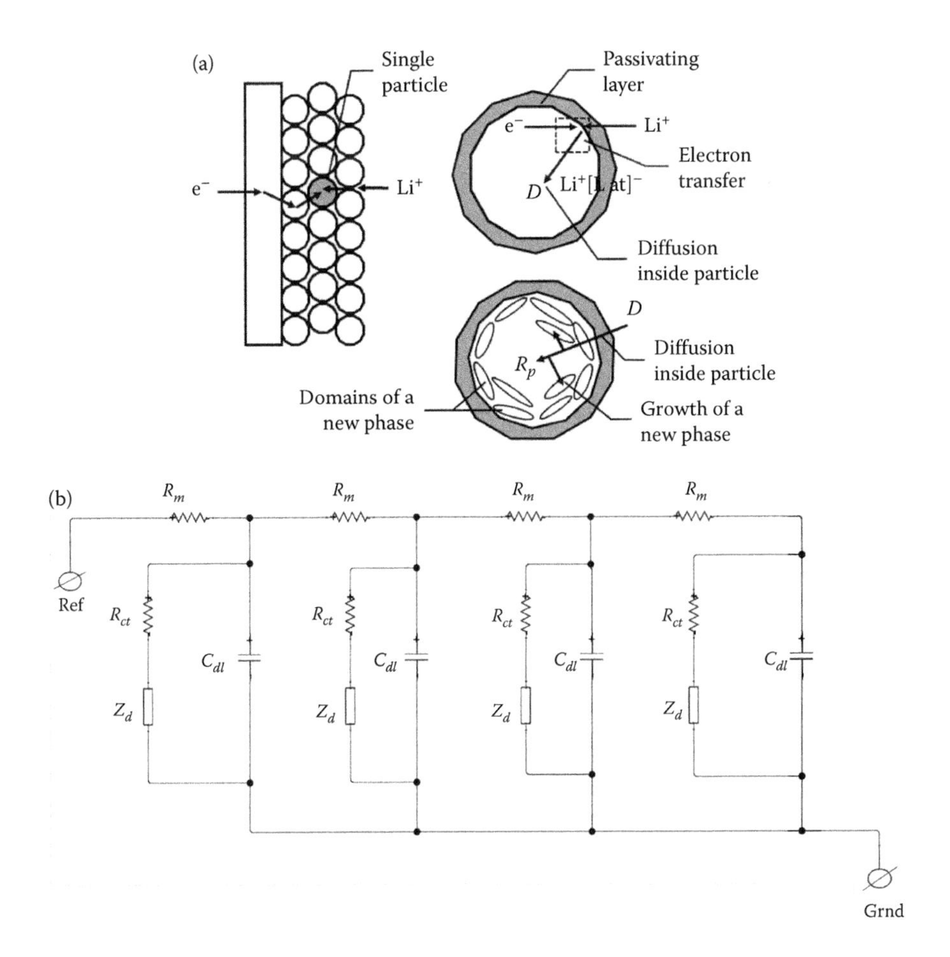

FIGURE 6.8 Kinetic model of Li-ion intercalation into mesoporous oxide materials proposed by Barsoukov et al. (a) and its equivalent circuit (b) [50].

$$Z_{(s)} = \sqrt{\frac{R_m}{s \cdot C_{dl} + \frac{1}{R_{ct} + Z_{d(s)}}}} \cdot \coth\left(\sqrt{R_m \cdot \left(s \cdot C_{dl} + \frac{1}{R_{ct} + Z_{d(s)}}\right)}\right)$$

$$Z_{d(s)} = \frac{\tanh\left(\sqrt{3R_d\left(s \cdot C_d + \frac{1}{R_p + C_p \cdot s}\right)}\right)}{\sqrt{\frac{3\left(s \cdot C_d + \frac{1}{R_p + C_p \cdot s}\right)}{R_d} - \frac{1}{R_d} \cdot \tanh\left(\sqrt{3R_d\left(s \cdot C_d + \frac{1}{R_p + C_p \cdot s}\right)}\right)}}, \tag{6.22}$$

where s is the imaginary angular frequency; R_m is the distributed resistance of transmission line representing electronic and ionic resistance of the layer of active material; Z_d is the impedance of infinite thin layer of active material including charge transfer resistance and passivation layer resistance on particle interface R_{ct} and double-layer capacitance C_{dl}; R_d and C_d are the resistance and capacitance of passivating layer on particles interface; R_p and C_p are the resistance and capacitance of a new phase formed inside particles.

So far, the kinetics of the intercalation of lithium into anode and cathode of LiBs has already been well analyzed by equivalent circuits based on different electrode/solution interfaces [41,49–55]. In order to take into account the geometry effects, grain-boundary effects, roughness, porosity, polycrystallinity, or particle-size distribution of electrode materials, the capacitor and the Warburg impedance are substituted by introducing constant phase elements (CPEs) into the *Randies equivalent circuit* [51,52]. Under the condition of the ac impedance measurements, the linearized Butler–Volmer relation applied, thus the partial Li-ion conductivity, the transfer coefficient and lithium-ion diffusion in the solid matrix of the electrode are well predicted by the proposed equivalent circuits.

6.3.2.1 Li-ion Diffusivity and Partial Conductivity

Kanoh et al. and Farcy et al. studied the Li-ion insertion/extraction reaction with a Pt/λ-MnO_2 electrode and a V_2O_5 electrode, respectively, by ac impedance spectroscopy [51,52]. Bojinov and coworkers propose the standard *Randies equivalent circuit* as shown in Figure 6.1c and d to analyze the overcharging process of a Li_xMnO_2 electrode, initially below 1.8 V, after 4 h at a current of 0.5 mA cm^{-2} and the low frequency Warburg diffusion impedance is also well predicted [53]. Ho and coworkers investigated the chemical diffusion, the component diffusion and the partial ionic conductivity of the cell Li|$LiAsF_6$(0.75 M) in propylene carbonate|Li_xWO_3, using the small signal ac impedance technique [54].

The chemical diffusion coefficient of Li-ion D_{Li} in the intercalation electrode materials can be evaluated from the ac impedance data. If the semi-infinite diffusion was satisfied in the ac impedance measurements, i.e., when $\omega \gg 2D_{Li}/L^2$, the chemical diffusion coefficient D_{Li} can be deduced from the slope σ of the plots Z_{Re} vs. $\omega^{-1/2}$ or $-Z_{Im}$ vs. $\omega^{-1/2}$ as shown in Eq. 6.17:

$$D_{Li} = \left(\frac{V_m}{\sqrt{2}nFS} \cdot \frac{dE}{dx} \cdot \frac{1}{\sigma} \right)^2, \tag{6.23}$$

where V_m is λ-MnO_2 molar volume, dE/dx the slope of the coulometric titration curve at each x value ($0 \le x \le 1$). If the finite length diffusion process occurs in the ac impedance measurements, a limiting resistance R_L is obtained ($R = R_\Omega + R_{ct} + R_L$), as shown in Figure 6.9.

In this case, the chemical diffusion coefficient D_{Li} can be deduced from the finite diffusion length L, the limiting resistance R_L and capacitance C_L, when ω

$$\begin{aligned} Z_{Re} &= R_L = -\frac{V_m}{nFS} \cdot \frac{dE}{dx} \cdot \frac{L}{3D_{Li}} \\ Z_{Im} &= \frac{1}{\omega C_L} = -\frac{V_m}{nFS} \cdot \frac{dE}{dx} \cdot \frac{1}{\omega L} \\ D_{Li} &= \frac{L^2}{3R_L C_L}. \end{aligned} \tag{6.24}$$

Besides the chemical diffusion coefficient, other kinetic quantities may also be derived as a function of the stoichiometry from the same measurements. For a predominantly electronic conducting sample, the partial Li-ion conductivity σ_{Li} may be deduced from the following equation:

$$\sigma_{Li} = \frac{nFD_{Li}}{V_m} \bigg/ \frac{dE}{dx}. \tag{6.25}$$

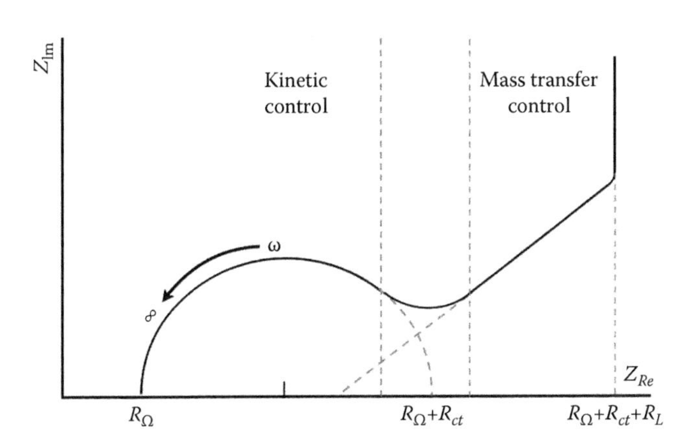

FIGURE 6.9 An impedance plot of the equivalent circuit shown in Figure 6.2 involving the finite diffusion effect.

6.3.2.2 Charge Transfer Resistance and Transfer Coefficient

The charge-transfer reaction occurring at the electrode/electrolyte interface in such as $LiTiS_2$ and $LiCoO_2$ rechargeable cells [55,56] can be summarized as follows:

$$Li^+ + Li_{1-x}M + e \rightleftharpoons Li_xM, \tag{6.26}$$

where Li_xM (M = TiS_2, CoO_2, etc.) denotes the Li-ion occupied site in the intercalation electrode M; $Li_{1-x}M$ is the empty activity site in the intercalation electrode M; Li^+ is the Li-ion in the electrolyte. Supposed the maximum intercalation concentration of Li-ion in the electrode material M is Γ (mol/m^3) and the concentration of Li-ion in the electrolyte near the electrode/electrolyte interface is $C_{Li}{}^+$(mol/m^3), the forward and the backward reaction rates of the intercalation process are expressed as the following equation:

$$\begin{aligned} r_f &= k_f C_{Li^+}(1-x)\Gamma \\ r_b &= k_b x\Gamma. \end{aligned} \tag{6.27}$$

Hence,

$$i = F\left(r_f - r_b\right) = F\Gamma\left(C_{Li^+}k_f(1-x) - k_b x\right). \tag{6.28}$$

Since intercalation of Li-ion into the electrode M results in the change of the intercalation reaction energy ΔG_{int}, we obtain

$$\begin{aligned} k_f &= k^0 \exp\left(\frac{-\alpha\left(F\left(E - E^0\right) + \Delta G_{int}\right)}{RT}\right) \\ k_b &= k^0 \exp\left(\frac{(1-\alpha)\left(F\left(E - E^0\right) + \Delta G_{int}\right)}{RT}\right), \end{aligned} \tag{6.29}$$

where k^0 is the standard rate constant of the intercalation reaction. Neglect the interaction energy of intercalated Li-ions and the empty positions, the Langmuir insertion isotherm is satisfied at equilibrium; hence, the total current i at equilibrium is zero and the exchange current i_0 is

$$\begin{aligned} i_0 &= Fk^0 C_{Li^+}(1-x)\Gamma \cdot \exp\left(\frac{-\alpha\left(F\left(E_{eq} - E^0\right) + \Delta G_{int}\right)}{RT}\right) \\ &= Fk^0 x\Gamma \cdot \exp\left(\frac{(1-\alpha)\left(F\left(E_{eq} - E^0\right) + \Delta G_{int}\right)}{RT}\right). \end{aligned} \tag{6.30}$$

Hence,

$$\frac{x}{1-x} = C_{\mathrm{Li}^+} \exp\left(\frac{-\left(F\left(E_{eq} - E^0\right) + \Delta G_{\mathrm{int}}\right)}{RT}\right). \tag{6.31}$$

Substitute Eq. 6.31 into Eq. 6.30 to yield

$$i_0 = Fk^0\Gamma C_{\mathrm{Li}^+}{}^{(1-\alpha)}(1-x)^{(1-\alpha)}x^{\alpha}, \tag{6.32}$$

where E^0 is the formal potential of the electrode. Consider the situation of high electrode polarization potentials, x approaches zero, which leads to a simplification of Eq. 6.31. That is,

$$x = C_{\mathrm{Li}^+} \exp\left(\frac{-\left(F\left(E_{eq} - E^0\right) + \Delta G_{\mathrm{int}}\right)}{RT}\right). \tag{6.33}$$

Substitute Eq. 6.32 into Eq. 6.10 to yield

$$R_{ct} = \frac{RT}{F^2k^0\Gamma C_{\mathrm{Li}^+}{}^{(1-\alpha)}(1-x)^{(1-\alpha)}x^{\alpha}}. \tag{6.34}$$

When x approaches zero, substitute Eq. 6.33 into Eq. 6.34 to yield

$$R_{ct} = \frac{RT}{F^2k^0\Gamma C_{\mathrm{Li}^+}} \cdot \exp\left(\frac{\alpha\left(F\left(E_{eq} - E^0\right) + \Delta G_{\mathrm{int}}\right)}{RT}\right). \tag{6.35}$$

Hence,

$$\ln R_{ct} = \left(\ln\left(\frac{RT}{F^2k^0\Gamma C_{\mathrm{Li}^+}}\right) + \frac{\alpha\Delta G_{\mathrm{int}}}{RT}\right) + \frac{\alpha F\left(E_{eq} - E^0\right)}{RT}. \tag{6.36}$$

As mentioned above, we note that the linear relation between $\ln R_{ct}$ and E_{eq} can be satisfied only in the limit of x approaching zero. From the logarithm of R_{ct} calculated from EIS measurements with various electrode potentials, the slop A of the linear relation of Eq. 6.36 can be obtained, which yields

$$\alpha = ART/F. \tag{6.37}$$

Zhang and coworkers studied the charge transfer resistance of the $LiCoO_2$ electrode with the increase of the electrode potential [56]. They observed a straight line at high potential region as shown in Figure 6.10, which is consistent with the prediction of Eq. 6.36. A value of 0.5 for α is calculated using Eq. 6.37.

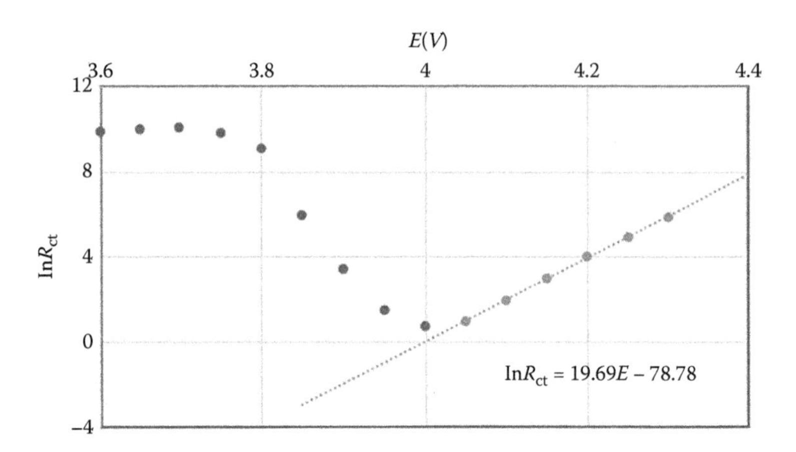

FIGURE 6.10 Variation of the logarithm of R_{ct} with the increase of the $LiCoO_2$ electrode potential [56].

REFERENCES

1. Dubal DP, Ayyad O, Ruiz V et al. Hybrid energy storage: The merging of battery and supercapacitor chemistries. *Chem. Soc. Rev*, 2015, 44(7):1777–1790.
2. Thomas KE, Newman J, Darling RM. in: WA van Schalkwijk, B Scrosati (Eds.), Chapter 12 Mathematical Modeling of Lithium Batteries, *Advances in Lithium-Ion Batteries*, Kluwer Academic/Plenum Publishers, New York, 2002, pp. 345–392.
3. Botte GG, Subramanian VR, White, RE. *Electrochim. Acta*, 2000, 45(15–16):2595–2609.
4. Gomadam PM, Weidner JW, Dougal RA, White RE. *J. Power Sources*, 2002, 110(2):267–284.
5. Arora P, White RE, Doyle M. *J. Electrochem. Soc.*, 1998, 145:3543–3553.
6. Santhanagopalan S, Guo Q, Ramadass P, White RE. *J. Power Sources*, 2006, 156(5):620–628.
7. Fuller TF, Doyle M, Newman J. *J. Electrochem. Soc.*, 1994, 141(1):1–10.
8. Doyle M, Newman J. *J. Appl. Electrochem.*, 1997, 27:846–856.
9. Sikha G, Popov BN, White RE. *J. Electrochem. Soc.*, 2004, 151(7):A1104–A1114.
10. Feng X, Gooi HB, Chen SX. "An improved lithium-ion battery model with temperature prediction considering entropy," in Proc. 3rd IEEE PES Int. Conf. Exhibition Innov. Smart Grid Technol. (ISGT Europe), Berlin, Germany, Oct. 14–17, 2012, pp. 1–8.
11. Taesic K, Wei Q. A hybrid battery model capable of capturing dynamic circuit characteristics and nonlinear capacity effects. *IEEE Trans. Energy Convers.*, 2011, 26(4):1172–1180.
12. Musolino V, Piegari L, Tironi E. New full-frequency-range supercapacitor model with easy identification procedure. *IEEE Trans. Ind. Electron.*, 2013, 60(1):112–120.
13. Zubieta L, Bonert R. Characterization of double layer capacitors for power electronics applications. *IEEE Trans. Ind. Appl.*, 2000, 36(1): 199–205.
14. Belhachemi F, Rael S, Davat B. "A physical based model of power electric double-layer supercapacitors," in Proc. IEEE Ind. Appl. Conf., Rome, Italy, Oct. 2000, vol. 5, pp. 3069–3076.
15. Venet P, Ding Z, Rojat G, Gualous H. Modelling of the supercapacitors during self discharge. *EPE J.*, 2007, 17(1):6–10.

16. Buller S, Karden E, Kok D, De Doncker RW. "Modeling the dynamic behavior of supercapacitors using impedance spectroscopy," in Proc. IEEE Ind. Appl. Conf., Chicago, IL, 2001, vol. 4, pp. 2500–2504.
17. Marie-Francoise JN, Gualous H, Berthon A. Supercapacitor thermal- and electrical-behaviour modelling using ANN. *IEEE Proc.: Electric Power Appl.*, 2006, 153(2):255–262.
18. Conway BE. *Electrochemical Supercapacitors: Scientific Fundamentals and Technological Applications*, Kluwer Academic/Plenum, New York, 1999.
19. Huang J, Sumpter BG, Meunier V. Theoretical model for nanoporous carbon supercapacitors. *Angew. Chem.*, 2008, 47(3):520.
20. Musolino V, Piegari L, Tironi E. New full-frequency-range supercapacitor model with easy identification procedure. *IEEE Trans. Ind. Electron.*, 2013, 60(1):112–120.
21. Manla E, Mandic G, Nasiri A. Testing and modeling of lithium-ion ultracapacitors. IEEE Energy Conversion Congress and Exposition. IEEE, 2011: 2957–2962.
22. Buller S, Thele M, Doncker RWAAD et al. Impedance-based simulation models of supercapacitors and Li-ion batteries for power electronic applications. *IEEE Trans. Ind. Appl.*, 2005, 41(3):742–747.
23. Dougal RA, Gao L, Liu S. Ultracapacitor model with automatic order selection and capacity scaling for dynamic system simulation. *J. Power Sources*, 2004, 126(1–2):250–257.
24. Gao DW, Mi C, Emadi A. Modeling and simulation of electric and hybrid vehicles. *Proc. IEEE*, 2007, 95:729–745.
25. Lukic SM, Cao J, Bansal RC, Rodriquez F, Emadi A. Energy storage systems for automotive applications, *IEEE Trans. Ind. Elec.*, 2008, 55(6):2258–2267.
26. Amatucci GG, Badway F, Du Pasquier A, Zheng T. *J. Electrochem. Soc.*, 2001, 148:A930–A939.
27. Simon P, Gogotsi Y, Dunn B. Where do batteries end and supercapacitors begin? *Science*, 2014, 343(6176):1210–1211.
28. Smith PH, Tran TN, Jiang TL et al. Lithium-ion capacitors: Electrochemical performance and thermal behavior. *J. Power Sources*, 2013, 243:982–992.
29. Shi L, Crow ML. Comparison of ultracapacitor electric circuit models. Power and Energy Society General Meeting – Conversion and Delivery of Electrical Energy in the 21st Century. IEEE, 2008: 1–6.
30. Wei T, Qi X, Qi Z. An improved ultracapacitor equivalent circuit model for the design of energy storage power systems. International Conference on Electrical Machines and Systems. IEEE, 2007: 69–73.
31. Buller S, Karden E, Kok D et al. Modeling the dynamic behavior of supercapacitors using impedance spectroscopy. *IEEE Trans. Ind. Appl.*, 2002, 38(6):1622–1626.
32. Wang Y, Carletta JE, Hartley TT et al. An ultracapacitor model derived using time-dependent current profiles. Symposium on Circuits & Systems. IEEE, 2008: 726–729.
33. Grama A, Grama L, Petreus D et al. Supercapacitor modelling using experimental measurements. Signals, Circuits and Systems, 2009. ISSCS 2009. International Symposium on. IEEE, 2009:1–4.
34. Srinivasan V. Mathematical modeling of electrochemical capacitors. *J. Electrochem. Soc.*, 1999, 146(5):1650–1658.
35. Bertrand N, Briat O, Vinassa JM et al. Porous electrode theory for ultracapacitor modelling and experimental validation. Vehicle Power and Propulsion Conference, 2008. VPPC '08. IEEE. IEEE, 2008:1–6.
36. Bertrand N, Sabatier J, Briat O. et al. Embedded fractional nonlinear supercapacitor model and its parametric estimation method. *IEEE Trans. Ind. Electron.*, 2010, 57(12):3991–4000.

37. Rao R, Vrudhula S, Rakhmatov DN. Battery modeling for energy-aware system design. *Computer*, 2015, 36(12):77–87.
38. Sluyters-Rehbach M, Timmer B, Sluyters JH. On the impedance of galvanic cells. *Zeitschrift Für Physikalische Chemie*, 1967, 52(1–4):89–103.
39. Sluyters-Rehbach M, Sluyters JH. On the impedance of galvanic cells: XXIX. The potential dependence of the faradaic parameters for electrode processes with coupled homogeneous chemical reactions. *J. Electroanal. Chem. Interfacial Electrochem.*, 1970, 26(2–3):237–257.
40. Sluyters JH. On the impedance of galvanic cells: I. Theory. *Recueil des Travaux Chimiques des Pays-Bas*, 2015, 79(10):1092–1100.
41. Boukamp BA. A package for impedance/admittance data analysis. *Solid State Ionics*, 2017, 18(1):136–140.
42. Bard AJ, Faulkner LR. *Electrochemical Methods, Fundamental and Applications*, John Wiley & Sons, New York, 2001, 99.
43. Shih H. Electrochemical Impedance Spectroscopy for Battery Research and Development, Technical Report 31, Solartron Instruments, 1996.
44. Mahon PJ, Paul GL, Keshishian SM, Vassallo AM. Measurement and modeling of the high-power performance of carbon-based supercapacitors. *J. Power Sources*, 2000, 91:68–76.
45. Miller J. in: S Wolsky, N Marincic, 4th International Seminar on Electrochemical Capacitors and Similar Energy Storage Devices, Florida Educational Seminars, Boca Raton, FL, 1994.
46. Pell WG, Conway BE, Adams WA et al. Electrochemical efficiency in multiple discharge/recharge cycling of supercapacitors in hybrid EV applications. *J. Power Sources*, 1999, 80(1–2):134–141.
47. Musolino V, Piegari L, Tironi E. New full-frequency-range supercapacitor model with easy identification procedure. *IEEE Trans. Ind. Electron.*, 2013, 60(1):112–120.
48. Tironi E, Musolino V. "Supercapacitor characterization in power electronic applications: Proposal of a new model," in Proc. Int. Conf. Clean Elect. Power, 2009, pp. 376–382.
49. Levi MD, Aurbach D. Simultaneous measurements and modeling of the electrochemical impedance and the cyclic voltammetric characteristics of graphite electrodes doped with lithium. *J. Phys. Chem. B*, 1997, 101(23):4630–4640.
50. Barsoukov E, Kim DH, Lee HS et al. Comparison of kinetic properties of $LiCoO_2$, and $LiTi_{0.05}Mg_{0.05}Ni_{0.7}Co_{0.2}O_2$, by impedance spectroscopy. *Solid State Ionics*, 2003, 161(1–2):19–29.
51. Kanoh H, Feng Q, Hirotsu T et al. AC impedance analysis for Li+ insertion of a Pt/MnO_2 electrode in an aqueous phase. *J. Electrochem. Soc.*, 1996, 143(8):2610–2615.
52. Farcy J, Messina R, Perichon J. Kinetic study of the lithium electroinsertion in V_2O_5 by impedance spectroscopy. *J. Electrochem. Soc.*, 1990, 137(5):1337.
53. Bojinov M, Geronov Y, Pistoia G, Pasquali M. *J. Electrochem. Soc.*, 1993, 140:294.
54. Ho C, Raistrick ID, Huggins RA. Application of AC techniques to the study of lithium diffusion in tungsten trioxide thin films. *J. Electrochem. Soc.*, 1980, 127(127):343–350.
55. Shen DH, Subbarao S, Nakamura BJ, Yen SPS, Bankston CP. Capacity decline of ambient temperature secondary lithium battery. Primary and Secondary Ambient Temperature Lithium Batteries, 1988, vol. 35, pp. C405.
56. Zhuang Q, Xu J, Fan X et al. $LiCoO_2$ electrode/electrolyte interface of li-ion batteries investigated by electrochemical impedance spectroscopy. *Sci. China Ser. B: Chem*, 2007, 50: 776–783.

7 Applications and Economics of Lithium-Ion Supercapacitors

Hongbin Zhao, Muhammad Arif Khan, and Jiujun Zhang
Shanghai University

CONTENTS

7.1 APPLICATIONS OF LITHIUM-ION SUPERCAPACITORS

7.1.1 Development Trend of Lithium-Ion Supercapacitors

A supercapacitor, also known as an ultracapacitor, is an electrical component having greater capacitance and power density than conventional capacitors. In the 2010s, the different product categories of supercapacitors show obvious change with the development of global renewable energy applications and vehicles. The traditional electronic double-layer capacitor (EDLCs) occupied the largest market share of 48.2%. However, the lithium-ion capacitor (LIC) segment is anticipated to witness the fastest growing segment. The global hybrid electric vehicle (HEV) market is growing rapidly. To reduce carbon emission, the hybrid supercapacitor is one of the major factors responsible for the exponential growth in the coming years. The market of hybrid supercapacitors is forecasted to witness the fastest growth with a rate of 23.5% during the forecast period.

Among the different types of supercapacitor, the board-mounted supercapacitor holds the largest market share in 2014 and is predicted to be the fastest growing market with a compound annual growth rate (CAGR) of 22.7% from 2015 to 2023. Increasing usage of board-mounted supercapacitors by different automotive manufacturers for various applications, such as regenerative breaking and start-stop system to reduce fuel consumption, is responsible for the robust growth of the board-mounted supercapacitor market.

Geographically, the Asia Pacific supercapacitor market is predicted to go through the fastest growth rate with a CAGR of 23.2% from 2015 to 2023. Rapid expansion of production facility in this region owing to huge capital investment by different companies in this emerging technology results in a sharp fall of the price of the supercapacitor. This reduction in the price of supercapacitors coupled with low maintenance cost is predicted to increase the usage of supercapacitors across different application segments. This, in turn, is expected to boost the demand for supercapacitor in the coming years.

The emergence of the supercapacitor fills the traditional gap between capacitors and batteries, and it is widely used in many fields, such as digital products, intelligent instruments, toys, power tools, new energy vehicles, new energy power generation system, distributed power systems, high power weapons, movement control, energy-saving construction, industrial energy conservation, and emission reduction. It has become an important product of low carbon economy.

7.1.1.1 High Voltage, High Capacity, and Safety

In recent years, in response to the fossil fuel depletion and an effort to prevent global warming, people took a variety of countermeasures. For the problem of fossil fuels, people began to import natural energy, such as solar and wind power. In terms of preventing global warming, some countermeasures are discussed and developed, such as changing vehicles with high CO_2 emissions with electric and auxiliary driving motor. However, these countermeasures have led to the emergence of new topics, such as the instability of the power system and the increase of power consumption. To solve these problems, storage components are indispensable.

Previous storage components have been developed with a lithium-ion rechargeable battery (LIB), but for different purposes, the output characteristics and cycle life of charge and discharge of LIBs are limited. Overcome the application limitation of LIBs, a LIC with high output and long life shows a promising future. In this chapter, LICs will be introduced for the hybrid car market, and its growing market of application solutions are mentioned.

FDK Corp. in Japan developed a LIC with high output power density and excellent charge/discharge cycling performance. It has been used in the fields of high voltage compensation device and load averaging of solar power generation. Besides, it also makes progress in cars with high output power, such as hybrid power automobile.

7.1.1.2 High-Voltage and Large-Capacitance Lithium-Ion Capacitors

LICs use activated carbon as the cathode, carbon materials as the anode, and organic lithium salt as the electrolyte. Like other double-layer capacitor, the cathode of a LIC also has similar charge storage mechanism. The anode has a similar mechanism with that of LIB anode consisting a typical REDOX reaction. By adding lithium-ion in the electrochemical systems, the working potential of the LIC can raise high up to 4 V, and the charge capacity of the anode also increased. The charge capacity of one unit can increase to two times that of a typical electric double-layer capacitor. Therefore, LICs show higher potential and larger capacity than that of EDLCs. For example, the energy density per unit volume of 10–50 Wh/L in LICs, which is much greater than that of an EDLC with a 2–8 Wh/L capacity. Although the energy density is lower than that of the LIB, the LIC has high output density and long lifetime. It also has two characteristics: excellent high temperature characteristics and lower self-discharge compared with an EDLC.

7.1.1.3 Safety due to the Difference of Cathodes

At present, there are three main requirements for storage devices. They are as follows: safety, long life time, and low cost. In these requirements, safety is the most important factor. Electro-storage devices are used to store energy, while unstable storage will lead to danger when energy density increases. At present, to improve safety of LIBs, people used separators and other insulated materials. However, the natural safety of energy storage principle is the most important.

The main difference between LIBs and LICs is cathodes. Usually, LIBs use lithium-based oxides as cathodes, while LICs use activated carbon as cathode. Lithium-based oxides not only contain abundant lithium, but they can also lead to fire–oxygen. Therefore, if a short circuit occurs within the cell for some reason, the heat caused by a short circuit will decompose the lithium oxide and further lead to the thermal decomposition of the whole unit, leading to severe heat effect.

Commercially, the positive electrode of LIC is activated carbon. When an internal short circuit occurs, activated carbon will react with the negative electrode, and the positive electrode and electrolyte will not react after that; therefore, it will be safe in principle. However, a serious reaction between cathode and electrolytes will cause severe decomposition of electrode materials, resulting in more serious heat effect.

7.1.1.4 High-Temperature Durability

For the lifetime, the storage components are expensive because more electrode materials are needed. Moreover, if we prolong the lifetime, it can reduce the frequency of replacing electrodes, reduce waste release, etc. and the environmental effect will be decreased. To reduce the degradation to achieve long life of LIBs, the charge and discharge range (charge and discharge depth) are shortened. The available capacity is essentially reduced. However, the main goal is to expand the charge and discharge depth and increase the lifetime. The charge and discharge principle of EDLCs is simple to adsorb or remove the ions in the electrolyte to keep a long life, but it is difficult to extend the life in actual conditions.

The main problem associated with storage devices is the increase in internal temperature, when operating for a long time. When it recharges repeatedly, the internal resistance causes the temperature to rise, which can greatly affect its life. Therefore, the high temperature durability is the necessary condition. The degradation caused by high temperature is mainly because of the oxidative decomposition of the cathode electrolyte. The higher the cathode potential, or the higher the environment temperature, the easier the decomposition of the electrolyte. If an EDLC decreases the positive potential, the unit voltage will also descend; therefore, it is not able to keep specific capacity. In contrast to that of an EDLC, the unit voltage of the LIC will not greatly descend even reducing the positive potential, which can make sure capacity with a stable level. Besides, the positive potential is far from the redox region, the durability of high temperature is excellent.

7.1.1.5 Integrated as Portable Module and Combined with Zn-Acid Batteries to Deduce Cost

To be economical is a very competitive factor to expand the market. However, the price of the storage material is not the only factor; besides, the setting environment and life of the storage system are also important factors. A large storage device is not a good product if it is cheap. Its long-term reliability is very important. Once it has an issue with quality, it will lose the market trust and will eventually cause great loss to the company. In practical application, we need a high voltage or output power, which can be achieved by combining many single units to get a complete module. The cost of module should be low to make it suitable for commercialization.

LIC cost can be deduced per the following: The single unit has high voltage, which can reduce the number of units; the high-temperature durability is excellent, the setting condition is more relaxed; cut management costs by simplifying the manufacture process. The higher the unit voltage, the lower the number of units to be used. For example, when the pack voltage is 300 V, it requires a 120 EDLC unit with a unit potential of 2.5 V, whereas only 80 LIC units with a unit potential of 3.8 V are needed.

Due to its high temperature characteristics, it can be used in a wide temperature range. As with LIB, the setting of places is limited when very tight temperature management is required. However, if the high temperature durability is excellent, the environmental temperature can be relaxed. Therefore, the higher degree of freedom of setting places can contribute to the cost reduction. Cost management refers to the cost of battery management system (BMS). The charging and discharging curves of LIBs can change dramatically with the current value and temperature environment, so the cost of BMS to manage the charging status will increase. The charge/discharge curve will not change significantly with the current value. This trend will not change with the temperature. FDK claimed to manage voltage to master the charging state, thus reducing the BMS cost. Even if the current value of the input and output changes dramatically, the slope of the unit will not change, so the charging state of the unit can be easily managed.

7.1.1.6 The Electricity Recycling Market Makes Up More than Half of Lithium-Ion Capacitor Market

Currently, the LIC has four main applications: (1) instant low compensation device and UPS (uninterruptible power supply) backup power market; (2) hybrid electric vehicle, crane and construction machinery, and other electric power regeneration markets; (3) the market for solar power and wind power generation equalization; and (4) electric-assisted markets, such as hybrids and copiers. In which, the largest market is the power generation market, which is expected to account for more than half of total market. However, with the expansion of the smart grid, the charge equalization function for solar power and wind power will also form a huge market.

7.1.1.7 The Instantaneous Low-Compensation Device

LICs have been adopted in the fields such as the transient compensation device and the solar energy load equalization. FKD claimed that their instantaneous low power-compensation device can supply power for more than 5 min, which can be used to power down the electricity that happens within a minute or less. Due to the small capacity, the maximum compensation of EDLCs can only compensate for the loss of the instantaneous voltage caused by lightning in the several microseconds. The LIC has a large capacity, which can be used to supply the voltage descend caused by power failure from the usual line to the standby line in several second.

The instantaneous low-compensation device is not set on each device, but is used to compensate the whole plant, thus reducing the management cost. Lead batteries are the mainstream of the instantaneous low-compensation device, but the leakage current of lead batteries is large, and it needs to be used to maintain the voltage, so it is expected to be replaced by LICs in the future.

7.1.1.8 Power Load Equalization Test of Lithium-Ion Capacitors on Island

As examples the application of solar power load equalization, 21st independent model system of new energy import verification project in Pingcheng, Okinawa Island (150 kW); Kitakaito (90 kW); and Taramajima Island (230 kW) adopted the LIC as hybrid energy storage and transition. Okinawa Electricity Company set up 230 kW of solar power generation equipment on Tarama-jima Island, to test the power load equalization of LICs.

7.1.1.9 Combine with Lead Batteries

With respect to the power supply to starter, the use of an LIC can replace lead accumulator to provide large power. If the lead battery is discharged repeatedly, it will accelerate deterioration. Therefore, by using LIC and lead batteries in parallel, the large power can be released from low resistance LICs, which can prevent the lead accumulator from being degraded by large output changes.

The lead accumulator is not changed greatly by the parallel lead accumulator and low resistant LIC; thus, the deterioration can be prevented. In addition, it is important to power the on-board electric vehicle when the engine stops and when the generator stops. The motor is usually driven by a belt driven by the engine to gain energy, so the generator is directly related to the combustion effect. Therefore, the construction of the generator to disengage the power source can realize the vehicle with excellent fuel efficiency. However, even if the generator is disconnected from the power source, the steering wheel and other electric goods will still require a great power. Hence, it is believed that LICs are also helpful in meeting the high-power demand.

7.1.1.10 Best Suited for Hybrid Cars

Previously, because of the shortage of EDLCs, the hybrid car mainly uses Ni-MH charging battery. The LIC has four times the energy density of EDLC, so it can be used in hybrid cars. As mentioned above, it is possible to greatly expand the charge and discharge depth. Nickel-hydride rechargeable batteries and the LIB will degrade if the charging and discharge depth is increased, so the charge and discharge depth is only around 40%. In other words, only 40% capacity is used. In this way, although the capacity is small, if 100% depth charge and discharge is taken, the LIC structure module also can realize the shape dimension and the weight of the rechargeable battery.

7.1.1.11 Management Advantages of Lithium-Ion Capacitors

Moreover, the LIC has advantages in the management of lifetime, charging state, safety, setting, and system miniaturization. The management and safety advantages of life-span and charging state are based on the LIC characteristics. In terms of setting freedom degrees, because the LIC can tolerate high temperature, it is not seriously limited by storage place. In addition, due to resistance to high temperature, it is not necessary to adopt the cooling mode with strict temperature management, so it can help to realize the miniaturization of the system.

With the development of electrification step by step, energy storage components will be developed in hybrid cars with high output usability and high capacity; high output will especially expand the market. Although it has not been fully realized, it is very important to improve the life management of the components. The most desired thing to avoid is the sudden unusable condition. To avoid this kind of situation, it is necessary to manage the storage device with high precision. The stable charge and discharge performance and charge/discharge curve will be regulated by the LIC per different conditions, so it can be called the best storage device. In addition to hybrid cars, it can be used in electric power steering and electric vehicle air conditioner. LIC modules can be used for a variety of purposes like requirement for future units, FDK plans to increase capacity and improve low temperature characteristic. To improve specific capacity, they plan to develop units with more than 5,000 F static capacity while maintaining high output to make the module miniaturized. In terms of improving the low temperature properties, the unit operating at −40°C will be developed. However, operating at low temperatures only requires for a shorter time at the first startup. Considering that the hybrid cars will open the heaters, which need to start the engine, we must consider up to what extent we can improve the low temperature properties. This may need to consider the relationship with the cost of materials.

7.1.2 Current Application of Lithium-Ion Supercapacitors

7.1.2.1 Wind Energy Conversion [1]

The electrical grid may experience disturbances due to a sudden reduction in the wind supply if there is a high penetration of wind power. These disturbances lead to a mismatch between the supply of the wind power and the demand in the grid, which could lead to a deviation from the normal operating frequency. In the United States, the normal frequency at which electrical equipment is designed to operate is 60 Hz. A deviation of the frequency from the normal designed frequency can damage equipment. A rapid drop in the frequency could cause tripping of generating units, shedding of loads, or even lead to a system collapse. This imbalance between generation and load can be reduced by using ESS, as the stored energy would be used as backup for sudden reduction in supply. These systems can promptly discharge to meet the load and stabilize the grid by providing short duration power necessary to maintain the grid frequency within a nominal range.

Frequency support requires power to be delivered for a very short duration. Advanced storage technologies such as flywheels, superconducting magnetic energy storage (SMES), and electrochemical capacitors (ECs) and batteries are suitable for this application because they respond instantaneously to frequent and unpredictable changes in wind.

Wind power is the conversion of wind energy into a useful form of energy, such as using wind turbines to make electricity. In relation to the use of LICs, this ElectroniCast study deals mostly with small-scale wind power, which is the name given to wind generation systems with the capacity to produce up to 100 kW of electrical power.

This application would benefit transmission operators, such as regional transmission operators or independent system operators who want to mitigate short-term fluctuations in frequency. This could benefit ratepayers because fast response storage provides cost and reliability benefits to them.

The future of the LIC market, despite exciting innovative devices driven by technological advances and ecological/energy-saving concerns, still face challenges in overcoming performance/price limitations and in attracting widespread consumption. The use of LICs in wind-power generation applications is increasing, initiating from government-based initiative schemes, then to commercial/business avenues, and eventually to the consumer level. The ElectroniCast market forecast of consumption is presented for two major end user categories: government/commercial and residential/non-specific. The government/commercial category is forecasted to maintain its market leadership role during the forecast period, with 91% of the worldwide market in 2011 and eventually falling to 73% as residential/other applications begin using the (new) solution. There are many different sources of energy (coal, nuclear, natural gas, hydro, oil); therefore, the use of wind turbine-based products, as well as other alternative or renewable energy solutions (naturally replenished) are constantly striving to serve the needs of consumers.

Ireland intends to achieve 40% renewable energy by 2020. This is an ambitious goal, considering most of this electricity comes from large-scale wind farms. The challenge is that it is an island grid, with only limited connection to the UK. Some winter nights, the Irish grid must take 75% of its electricity from renewable sources. This calls for additional services, so Freqcon company deployed Ireland's first combined ultracapacitor and energy storage facility for the Tallaght Smart Grid Test bed in South Dublin County. These 300 kW/150 kWh systems were developed to demonstrate that a combination of lithium-ion batteries, Maxwell Technologies ultracapacitors, and Freqcon power converters can deliver what is needed. Though this is a trial run in Ireland, one of Freqcon's recent contracts was a 5 MW/10 MWh facility in China. Around 90% of its business is in Asia mainly in China and South Korea. One of its clients is China's leading wind turbine developer, Goldwind. Freqcon was one of Maxwell Technologies first customers in 2000. It first used the San Diego company's ultracapacitors for pitch control in wind turbines. Ultracapacitors respond much faster than battery storage, discharging in fractions of a second and with more immediate power. This makes them invaluable in grid applications like Ireland, where large amounts of wind energy are being fed into the grid. However, batteries produce and store energy through a chemical reaction, rather than storing it in an electric field, which means they have more capacity.

Two complementary technologies were combined to provide a service that neither would be able to do as efficiently and by combining LICs and batteries with a very cost efficient fast frequency response system, its advantages for wind energy are obvious, and it also can provide back-up power with the batteries.

7.1.2.2 Renewable Energy Generation Systems/ Distributing Electrical Power Systems

In renewable energy power generation or distributed power system, the power output of the power equipment has the characteristics of the instability and unpredictability.

A supercapacitor combined with a solar cell can be used in street lamps, traffic warning signs, and signposts. It can also be used in wind generation, full cell and other distributed generation systems, compensate large pulse current and as a backup power for electric cutting to improve the stability and reliability of power system. Supercapacitor energy storage is proposed, which can fully play their advantageous role, including large power density, long cycle life, high energy storage density, no maintenance, etc. In addition to storing energy, it can also be mixed with other energy storage device.

7.1.2.3 The Energy Buffer of the Variable Frequency Drive System

Supercapacitors and power converters constitute energy buffers, which can be used for elevators such as variable frequency drive system. When the elevator rises, the energy buffer supply for the drive system by DC power and provide the necessary motor peak power; When an elevator slows down, it can absorb the energy of the motor to the DC bus through the inverter feedback.

7.1.2.4 Military Equipment Field

Military equipment, especially the field equipment, mostly cannot use the public power directly, and need the power equipment and energy storage device. The requirements of the energy storage unit for military equipment are reliable, lightweight, and strong concealment. Supercapacitor and battery hybrid energy storage is proposed, which can greatly reduce the weight of radio and other equipment; solve low-temperature start difficult problem of vehicle, such as military vehicle, tanks and armored vehicles, improve the vehicle's power performance and concealment, solve the failure and short service life of battery for conventional submarines. It can also provide peak power for radar, communication, and electronic countermeasures system, and reduce the power level of the main power supply.

7.1.2.5 Application of Lithium-Ion Capacitor in Railway

Maxwell Technologies [2], as a developer and manufacturer of ultracapacitor-based energy storage and power delivery solutions, announced the first commercial application of LICs, developed in conjunction with China Railway Rolling Stock Corporation (CRRC-SRI), China's largest rail manufacturer. The technology will be used for rapid energy regeneration in the trolley system in the capital city of the Hunan province in China. Following last year's announcement of Maxwell Technologies strategic partnership with CRRC-SRI to collaborate on developing next-generation capacitive energy storage solutions, this project is the first to leverage Maxwell Technologies new lithium-ion technology and validates its unique value proposition for rail applications.

CRRC-SRI will leverage LICs for the Changsha Subway in Changsha, Hunan, as the single source of power for instant charging and discharging to propel the trolley. Lithium-ion capacitors can charge light rail vehicles in 30 s and keep them going for 5–10 min, ensuring the trolley will be able to restart quickly in constant stop-and-go traffic. The technology fulfills China's rail requirements for energy savings and environmental protection.

Maxwell Technologies' LICs combine ultracapacitors' high power density with lithium-ion batteries' high energy density for onboard ESS. Compared to traditional ultracapacitors, LICs got triple energy density and reduce the total weight of the energy storage system by 50%. Maxwell Technologies' LICs fill that market demand today and ensure our trolleys are constantly moving. The commercial use of LICs represents a transformation in the energy storage industry and a significant opportunity for Maxwell Technologies to revolutionize the way power is distributed in rail applications in China. In this case, Maxwell Technologies' LICs are the perfect solution for the growing rail market. As demand for energy storage solutions increases, we believe there is a large opportunity to use the technology in even more applications, such as grid firming, wind pitch control and robotics, to drive future revenue growth.

7.1.2.6 Solar Combined with Lithium-Ion Capacitors

It is well known that solar cells can merely work under sunlight illumination; therefore, the output power of solar cells is unstable and intermitted. In this context, appropriate solar energy storage devices are particularly necessary. Today, different solar energy storage technologies, including pumped hydro, flywheel, compressed air, SMES, and various electrochemical technologies (lead-acid batteries, lithium-ion batteries, fuel cells, flow batteries, and supercapacitors, etc.), have been developed/or studied [3]. Among these, supercapacitors, also known as electrochemical capacitors, are regarded as one of the most promising energy storage devices due to their fast charge/discharge rates, high power density, long cycle life, environmental inertness, and an almost maintenance-free essence [4–9].

It has been reported that a solar cell can be connected with a supercapacitor through an external circuit to form an energy conversion and storage system [10–13]. Typical application is on street lamps, traffic lights and warning sign lamps.

However, such integrated systems via external connection have several problems. First, the use of an external connection circuit will reduce the energy storage efficiency (ESE) of the supercapacitor part [13]. In other words, external electrical wires with internal resistance will consume part of the power, resulting in the decrease of ESE. Most importantly, traditional energy conversion and storage systems are both large and bulky, and thus, can no longer meet the demands of portable and wearable electronic devices [14,15]. Therefore, to solve the above-mentioned problems, a new device with the dual function of photoelectric conversion (PC) and energy storage (ES) came into being. Such dual-functional devices integrating solar cells and supercapacitors are named as photo-supercapacitors (PSs) [13,16]. PSs are somewhat of an *in situ* energy storage device, where a solar cell and a supercapacitor are integrated into a single device by sharing a common electrode. In 2004, Miyasaka et al. integrated a dye-sensitized solar cell and a supercapacitor into a photo-supercapacitor for the first time [17]. And over the past 13 years of research, PSs have been developing rapidly. According to the type of division, dye-sensitized, polymer, quantum dots sensitized, and perovskite solar cell-type PSs have been reported one after another (Figure 7.1). And these research efforts mainly focused on demonstrating the feasibility of integrating the different types of solar cells with supercapacitors into dual-functional photo-supercapacitor devices.

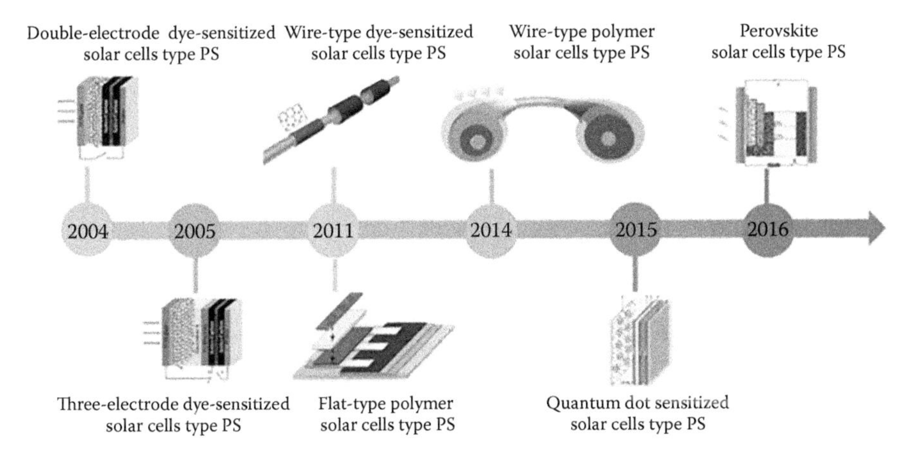

FIGURE 7.1 A chronology of the development of PSs. Images reproduced with permission as follow: "Double-electrode dye-sensitized solar cells type photo-supercapacitor": Reproduced with permission [17] Copyright 2004, American Institute of Physics. "Three-electrode dye-sensitized solar cells type photo-supercapacitor": Reproduced with permission [18]. Copyright 2005, The Royal Society of Chemistry. "Wire-type dye-sensitized solar cells type photo-supercapacitor": Reproduced with permission [19]. Copyright 2011, Wiley-VCH. "Flat-type polymer solar cells type photo-supercapacitor": Reproduced with permission [13]. Copyright 2011, The Royal Society of Chemistry. "Wire-type polymer solar cells type photo-supercapacitor": Reproduced with permission [20]. Copyright 2014, Wiley-VCH. "Quantum dot sensitized solar cells type photo-supercapacitor": Reproduced with permission [21]. Copyright 2015, Elsevier. "Perovskite solar cells type photo-supercapacitor": Reproduced with permission [22]. Copyright 2016, Wiley-VCH.

Since PSs are still in the early stages of development, they inevitably have some problems and challenges, including low charging voltage, low photoelectric conversion efficiency (PCE), and serious self-discharging, as well nonstandardized evaluation and difficulty of encapsulation [3–17,19,21,23–33].

Low charging voltage: In the ideal state, the maximum charging voltage of the ES part in a PS should be equal to the V_{oc} of the PC part. However, this voltage is less than that. The reasons for this remain unclear and need to be researched vigorously in the future. Moreover, most DSSCPSs typically use the liquid-phase I^{-}/I^{3-} redox electrolytes in the PC part, allowing for a maximum V_{oc} of only 0.7−0.85 V [18]. Thus, the maximum charging voltage of all reported DSSCPSs can reach a maximum voltage of 0.8 V [18]. On the other hand, from the energy density formula of supercapacitors ($E = 1/2\ CV^2$), we know that the charging voltage largely determines the energy density of PSs. Thus, low charging voltage leads to low energy density of PSs [14].

Low PCE: Generally, the PCE of PSs is far less than the maximum PCE of the same type of solar cells. Compared to the maximum certified PCE of 22.1% for perovskite solar cells [34–36], this PCE must be increased for future applications.

Serious self-discharging: Self-discharging is a very serious problem of PSs, which will undoubtedly hinder their future applications. The reasons for self-discharging can possibly be the following: (1) the ES part itself has a certain internal resistance, which will consume energy [37]; (2) during the discharge process, the electrons of the PC part will flow back to the cathode of the ES part, and the recombination between the electrons and the positive charges will occur, leading to the reduction in the stored charges [18,28].

Difficulty of encapsulation: Compared with flat-type PSs, the encapsulation of the wire-type PSs is more difficult, especially for devices using liquid-phase electrolytes. Moreover, organic halide perovskite solar cells are vulnerable to moisture, which results in a reduction of the device performance. Thus, ensuring proper encapsulation of wire-type PSs without affecting their performance and improving the stability of the PSs utilizing perovskite solar cells as their PC part remains a major challenge.

Non-standardized evaluation on the performances for PSs: Currently, there is no standardized method to evaluate the performance of the PSs, although the inherent performance parameters of the traditional solar cells and supercapacitors coupled with OPCESE values have been used as evaluation parameters for PSs. In the current stage, researchers generally focus on a few of the performances of a single PC or ES part, ignoring the overall performance of the entire device. Therefore, this method is inaccurate and makes it impossible to objectively compare the different types of PSs.

7.1.3 Industry Developments and Problems of Lithium-Ion Supercapacitors

LICs have the unique characteristics of high energy density and high power density; their raw material requirements are relatively harsh. Electrode, current collector, electrolyte, and membrane composition and quality have a decisive impact on the performance of LICs [38,39].

Upstream industry: The upstream stage of the production process involves searching for and extracting raw materials, e.g., electrode, electrolyte, separator, current collectors and metal lithium. Midstream industry: single cell with various shapes and specification, integrated system modules. Downstream industry: The downstream stage in the production process involves processing the module collected during the midstream stage into a finished product. The downstream stage further includes the actual sale of that product to other businesses, governments, or private individuals. Downstream is the application of end-market demand; the Japanese market is preliminary opening, in the local market, then further flourishing in the international market because of its widespread application in different devices, e.g., Wind power, LED lighting, solar, hybrid electric vehicles, etc.

It has been only a few years ago, since LICs were first developed, but the products are still being tested in factory and the market is not widely open due to the high cost of technology. In practical production, LIC technology is much more complex than that of LIBs and EDLCs, and the current research focuses mainly on the two aspects, the carbon negative electrode pre-lithiation technology and system.

In 2000, Morimoto et al. reported a hybrid capacitor used in lithium graphite cathode and lithium electrolyte, and its working voltage range can reach 3–4.2 V. Subsequently, the samples they assembled has specific capacity of 16 Wh/L, specific power density of 500 W/L, and 40% decay rate after 40,000 cycles.

2005, Fuji heavy industries has disclosed LIC manufacturing technology in succession. Its negative electrode is embedded in a large number of lithium-ions and the benzene material, positively activated carbon. In 2006, Fuji Heavy's Hatozaki reported the lithiated carbon material/activated carbon LIC, its working voltage is 2.2–3.8 V, which can reach 12–30 Wh/kg, and the capacity retention rate is above 96% after 300,000 cycles. He said that the LIC had such excellent charge and discharge properties, thanks to the pre-lithiated carbon negative material.

After Fuji, Heavy Industry Corp. claimed the LIC manufacturing technology, Japan and other countries' scientific research institutions and enterprises also began to pay attention to the hybrid energy storage technology, began to develop LICs, and tried to launch their own products. Japanese companies lead the industrialization of LICs (Table 7.1).

7.1.3.1 The Problems Need to be Solved in the Industrialization of Lithium-Ion Capacitor

Research history of LICs is not long; the current products from several Japanese companies gradually show good performance, and demonstrated applications on wind power generation, in short-term power compensation device, hybrid industrial areas and micro-mechanical vehicle.

From the aspect of industrialization, to further improve the energy density, besides pre-lithiation technology, electrode materials and system matching research, there is a lot of work needed to be carried out, especially in the development of high-performance electrode materials, the electrolyte matching, monomer production process optimization, and unit detection method.

7.1.3.2 Carbon Cathode Pre-lithiation Technology

Pre-lithiation technology is one of the most important parts, and, its high manufacturing cost and complex process are recognized technical difficulties. The existing data have revealed the various production techniques of LICs. The selection of lithium sources, the lithiation process, and lithium doping determines the reliability of the device performance manufacturing cost.

TABLE 7.1
Product Parameters of Japanese LIC Companies

Company	Voltage (V)	Capacity (F)	DC Impedance (mΩ)	Working Temperature (°C)	Specific Energy Density (Wh/kg)	Specific Energy Density (Wh/L)	Cycling Life
JM Energy	2.2–3.8	3,300	1.0	−30 to 70	12	20	100,000
AFEC	2.0–4.0	~2,000	~1.	−25 to 80	–	18	100,000
Taiyo Yuden	2.2–3.8	200	~50 AC	−25 to 70	10	20	100,000
New Kobe motor	2.2–3.8	1,000	3.5	−15 to 80	–	10	–

Fuji Heavy industry uses porous metal foil as a collector, placing a piece of lithium foil in the opposite position of the outermost cathode. In this way, even a monomer containing a multilayer electrode, Li^+ is free to move through the layers of electrodes and attaching on the collector; further Li ions can be easily doped in to anode. The technology is based on lithium foil as the supply source of Li, but the soft lithium foil and strict environmental requirements make the assembly of the monomers extremely inconvenient and accompanied by greater security risks. The Samsung Motor deposited lithium film on the surface of one side of separator by vacuum PVD, which made the lithium film and the negative electrode face to face, and the Li^+ in the lithium film was doped in anode [40]. Compared to Fushi Heavy Industry Corp., Samsung Motor's method has the following advantages: (1) due to the lithium film in directly contacts with an anode, in the subsequent pre-lithiation process, porous collector is not needed; therefore, internal resistance will decrease; (2) this method can control the usage amount of lithium conveniently, and the safety will be enhanced; (3) each anode layer contact with lithium foil directly, which can greatly shorten pre-lithiation time. However, the practical feasibility of this method remains to be proven.

Zheng's group [41] used metal lithium powder (SLMP) as the source of lithium resource. The SLMP has particle size of 10–200 nm, and passivation treatment was done for the surface. For the passivation, the dry-method was used to get an electrode after mixed with hard charcoal. With activated carbon as an anode and SLMP as a cathode, respectively, the LIC monomer was assembled. The test results showed that the specific energy of the monomers was about 25 Wh/kg, and the discharge specific energy of 44 C was about 60% of 2.4 C. The capacity decay rate is only 3% after 600 cycles. Compared with Fuji heavy industry, which uses lithium foil structure, this LIC can be made in the drying room, without the harsh environment of the glove box, which greatly increases the operability [42].

7.1.3.3 Research on the Doping Amount of Lithium

The doping of lithium is a key parameter in the pre-lithiation process. With an excessive amount of doping, the product will have residual lithium source after pre-lithiation, which will affect the product capacity and cause safety problem. If the amount of doping is too low, then the improvement of the voltage and energy density will not meet the requirement. Therefore, it is necessary to design rational lithium doping structure to make safe and reliable products.

Kumagai et al. [43] tested the properties of graphite with different pre-lithiation degree. The results show that the pre-lithiation degree determines the ratio of capacitance and cycle performance in the initial stage of the monomer, 70% of pre-lithiation degree is suitable for the LIC with high rate capacitance and stable cycling performance. Ping et al. [44] used graphited intermediate carbon microsphere (MCMB) as anode, activated carbon as cathode to assemble LIC. The XRD pattern of the LIC anode (LMCMB) with pre-lithiation amount of 50, 150, 200, 250, and 300 mAh/g was studied. The results showed that when the lithium amount was less than 200 mAh/g, the LMCMB was able to maintain a good graphite crystal structure, while the structure of the crystal was destroyed when the lithium content was higher than 250 mAh/g. Zhang [45] also validated this conclusion.

7.1.3.4 Research on Pre-lithiation Method

Instability of pre-lithiated materials in the loop are the source causes of the product capacity decay, and the method of pre-lithiation is crucial. The reasonable pre-lithiation method can guarantee the stability of the lithiated anode materials to ensure that the cycle stability of the cell unit.

The Australian energy agency (CSIRO) has conducted a thorough study of the issue [46]. They made Li/graphite/AC three electrode LIC, and studied the pre-assigned lithium effect of graphite by the following three methods: to short-circuit the Li/graphite electrode and directly discharge to insert lithium in graphite; the graphite electrodes were charged at 0.05 C constant current to insert lithium; inserting lithium in graphite electrode by recharging and discharging cycle with the external resistor. Check the discharge status of the battery by monitoring the lithium/predoped graphite potential and the open circuit voltage. Results show that with the first kind of method for pre-lithiation about 10 h, the Li doping is about 71% of the theoretical capacity of graphite. With the cycling test, doped lithium in the graphite electrode is lost, and the self-discharge phenomenon of single cell is more serious, thus inferring that this approach fails to produce uniform SEI film; the self-discharging of the second method is a bit improved compared with the first one. The third kind of method showed very low self-discharging, indicating uniform SEI membrane formed on the surface of graphite.

Yuan et al. [47] have also made the study of the three different pre-lithiation ways, i.e., the short circuit, constant current, and charge/discharge cycling. The results show that the performance of the device can be affected by high or low lithium insertion potential. In the case of constant current, the lower the lithium insertion potential, the larger the insertion rate. However, low lithium insertion rate leads to negative side reaction, which affects C-rate performance of the single cell. Compared with simple constant current, the electrochemical performance significantly improved when the negative electrode suffered cycling at constant current.

Decaux [48] developed a new method of pre-lithiation. Graphite electrode and lithium electrode were assembled into a two-electrode battery with 2 mol/L of LiTFSI in DOL-DME as electrolyte. After 10 cycles of charges/self-discharge pulse, the anode became pre-inserted lithium of graphite. Then, the pre-treated graphite electrode and activated carbon electrode were assembled to the LIC. Its work potential range is from 1.5 to 4.2 V, and the capacity is 60% higher than that of pure activated carbon, the energy density is up to 80 Wh/kg.

7.1.3.5 Study on Electrode Material and System Matching

The properties of electrolyte electrode materials and their matching are of vital importance to the electrochemical properties of energy storage devices. Energy store procedure of LICs contains both body phase reversible lithium-ion insertion/deinsertion REDOX reaction in the graphite electrode materials, and physical adsorption of the charge on activated carbon wit high specific surface. Based on the short board effect, energy characteristics of LIC depends on activated carbon electrodes, power characteristic depends on battery materials such as graphite electrodes. LIC can achieve high energy density and power density by the match of negative electrode materials and electrolyte.

Wang used five different types of graphite as the anode and AC as the cathode, studied the effect of different mass ratio of graphite/AC and two different electrolytes ($TEMABF_4$-PC and $TEMAPF_6$-PC) on the performance of the capacitor [49,50]. The study found that the capacitance of the unit was more dependent on the positive negative than the graphite type. With the increase of graphite/AC mass ratio, the unit capacitance increased significantly, but the cycling performance became worse. For LIC with the same graphite/AC mass ratio, the unit capacity of $TEMABF_4$-PC electrolyte was significantly higher than that of the $TEMAPF_6$-PC electrolyte.

Csiro continued to study the LIC properties with graphite as anode. Studying LIC performance using the seven kinds of commercial graphite as cathode, they found that by reducing the coating thickness of graphite electrode, the C-rate performance of unit can be improved. Graphite with smaller particle size can improve the performance of C-rate, while leading to the irreversible capacity loss [51,52]; After graphite being ball-milled processing, low rate (0.1 C) charge and discharge, the discharge capacity of ball-milled graphite is lower than the original graphite. At high ratio (1 C–60 C) charge and discharge, discharge properties and cycle performance of ball-milled graphite are better than the original graphite [52]; In the voltage range of 3.1–4.1 V, the unit discharge capacity is 55 Wh/kg, and the capacity retention rate is 97% after 100 cycles.

Schroeder et al. [53] produced the LIC of the cathode material with soft carbon petroleum coke for the poor performance of graphite at large current density. The test results showed that the specific energy and specific power of units were 48 Wh/kg and 9 kW/kg, respectively, and the cycle life was 50,000 times. Karthikeyan et al. [54] studied LIC with Li_2FeSiO_4 (LFSO) as the cathode, AC as the anode and 1 mol/L $LiPF_6$-EC/DMC as the electrolyte. CV test results show that this LFSO/AC system has the capacitance characteristic in voltage of 0–3 V. Under the current density of 1 mA/cm^2, the maximum discharge capacity of the LIC is 49 F/g, and the energy density and power density is 43 Wh/kg and 200 W/kg, respectively.

The energy that the cathode can store determines the energy density of the battery unit; therefore, optimized suitable intercalated-Li anode materials and developed anode material with high capacity are also particularly important [55]. Li et al. [56] added the graphene (GO) in the AC cathode to get a composite electrode. By the functionalization of GO surface, they effectively increased the number of adsorbed sites or reaction sites. Under the power density of 2 kW/kg, the energy density of the AC-GO/prelithiation graphite capacitor is as high as 82 Wh/kg, and the capacity of the hybrid is basically stable after 1,000 cycles.

In the present study, LICs generally use organic electrolyte of traditional LIBs or EDLCs directly, such as Cao [57,58] used the $LiPF_6$/EC DEC PC electrolyte, Sivakkumar etc. [59,60] used the $LiPF_6$/EC DMC EMC electrolyte, Schroeder, etc. [61] used the $LiPF_6$/EC DMC electrolyte, and Wang [62,63] used the $TEMABF_4$/PC or $TEMAPF_6$/PC electrolyte. LICs need a lot of charge carriers in electrolyte (ions) to compensate for charge/discharge; therefore, an EMITFSI LiTFSI–TFEP with ionic liquid, lithium salt, and fluorinated alkyl phosphate ester was developed [64].

7.2 ECONOMICS OF LITHIUM-ION SUPERCAPACITORS

The global supercapacitor market is showing a significant potential and is projected to witness compound annual growth rate of 18.60% over the forecast period to reach $2.44 billion in market size by 2020. Rapid technological advancements and quickly evolving green energy applications have proved to be the major drivers for the market. The transportation industry has given the major boost to the supercapacitor market since the demand for electric vehicles and trains are increasing, especially in the developed economies like China, the United States, Japan, France and others. Due to the features, like regenerative braking and easy applications in hybrid vehicles, supercapacitors have become useful to transportation and industrial applications.

Supercapacitors for renewable energy application have grown over the period. Thus, increasing focus on renewable energy sources is a huge opportunity for the supercapacitor market. Major research and developments of supercapacitor, its variants and the potential materials are being done in the United States. Whereas, high prices and lack of wide industrial experience are the key impediments hindering the growth of the market.

North America held the largest market share in the global market and was projected to dominate throughout the forecast period, closely followed by Europe. The region is majorly driven by the United States owing increasing R&D happening in the region along with the strict regulation pertaining to environment friendly technologies existing in the region.

Some of the key players in the global supercapacitor market are Maxwell Technologies Inc., Panasonic Corporation, Skeleton Technologies, Cap-XX LTD, Skeleton Technologies, Graphene Laboratories INC., AVX Corporation, and several China companies.

7.2.1 Global Commercial Companies for Lithium-Ion Supercapacitors

2016 Global Lithium-Ion Capacitors Industry Report is a professional and in-depth research report on the world's major regional market conditions of the Lithium-Ion Capacitors industry, focusing on the main regions (North America, Europe, and Asia) and the main countries (the United States, Germany, Japan, and China) (Tables 7.2 and 7.3).

7.2.2 Economic Evaluation

Here we compared the cost advantage and disadvantage of several traditional and novel ESS, including pumped hydro systems (PHS) and Compressed Air Energy Storage (CAES) systems; batteries, including lead-acid (Pb-acid), sodium-sulfur (NaS), nickel-cadmium (Ni-Cd), lithium-ion (Li-ion), flow batteries including vanadium redox battery (VRB) and zinc bromine (ZnBr), SMES, EC, and flywheels.

TABLE 7.2
Some Supercapacitor Manufacturers in China

Company	Product Type	Advantage
Shanghai Aowei Tech	Gold capacitor, double-layer capacitor	Ultracapacitors for vehicles area leader
Beijing Hezhong Huineng Tech	HCC series high voltage-organic	Products are widely used in the automotive, energy and other areas • High power density, up to 300–5,000 W/kg, equivalent to 5–10 times the battery • Temperature range −40°C to +70°C • Capacity range is usually 0.1 F–10,000 F.
Beijing Star Tech	Jixing series	Can be customized according to user needs
Harbin Ju Rong new energy	VCT Light electric cars use power Electric forklift uses of power, VCS, hybrid cars, elevators use Power balance system uses power VCHDC screen, EPS, ECT Electric tricycle capacitor battery	Independent research, development and production of national patent products
Jinzhou Kamei Energy	Winding-type 2.5/2.7 V, combination type 5.0/5.5 V, laminated type 5.5 V	Sufficient production
Hangzhou Fukai Tech	Gold capacitor, double-layer capacitor	Products used in wireless communication, mobile computers, automobiles and other
Shandong Hite	Gold capacitor, double-layer capacitor	3,000,000 per month
Baina electric	2.3, 2.5, 2.7, 3.0 V	To meet the 2,000-MW class wind turbine production capacity, 3,000 new energy vehicles and other industrial power market ultracapacitor module requirements
Beijing Jiexikangke Trade Ltd	Gold capacitor, double-layer capacitor	Leverage industry resources to provide reasonable price product

In Figure 7.2, CAES system has the lowest storage system cost for load-shifting and frequency support. However, limitations due to active site availability and difficulty of excavation reduce its wide-spread applicability. PHS faces a similar challenge associated with geographical constraints due to the need for two reservoirs separated by an elevation. Although the calculated costs are very low, these storage systems may have other limitations that prevent them from being the best choice in many situations.

Pb-acid batteries have the lowest cost among the batteries for load shifting and frequency support and are very close for power quality. Pb-acid has relatively high energy costs, and is therefore most expensive for load shifting among the three application areas. NaS batteries, on the other hand, have a lower energy cost (relative to their power cost), and thus load shifting is the least costly among the three

TABLE 7.3
Main Production Enterprises in Global

Country	Company
USA	Maxwell
	Evans
	Skeleton
	Pinnacle
	Los Alamos National Lab
	US Army Fort Monmouth
	Tecate Group
Russia	Econd
	Esma
	Elit
German	Epcos
	Sma
Japan	NEC
	Panasonic
	Tokin
	Elna
French	Saft
	Alsthom Alcatel
	Bollore
Italy	Fiat
	Magneti Maralli SEPA
South Korea	Ness cap
	Korchip
	Nuintek
Northern Europe	Superfarad
Denmark	Danioncs
Australia	Capxx
New Zealand	Powerstor

applications. The high-power cost for NaS is driven by the need to maintain high operating temperatures.

Flywheel technology provides continuous power from a pulsating output by converting kinetic energy to electrical energy. Power quality requires short bursts of energy at high power for small durations. The nature of cyclic rapid recharges and discharges as well as lower cost suggests that flywheel technologies are well suited for power quality and frequency regulation. Load-shifting is a less attractive application for flywheels, as their size increases rapidly with the energy requirement.

The two flow batteries that we consider, VRB and ZnBr, have slightly higher costs for power quality among the three application areas. This is because they require expensive power conversion equipment and the need for extra auxiliary systems drive up the power cost. On the other hand, the low costs of zinc metal and their ready availability makes ZnBr suitable for long duration application.

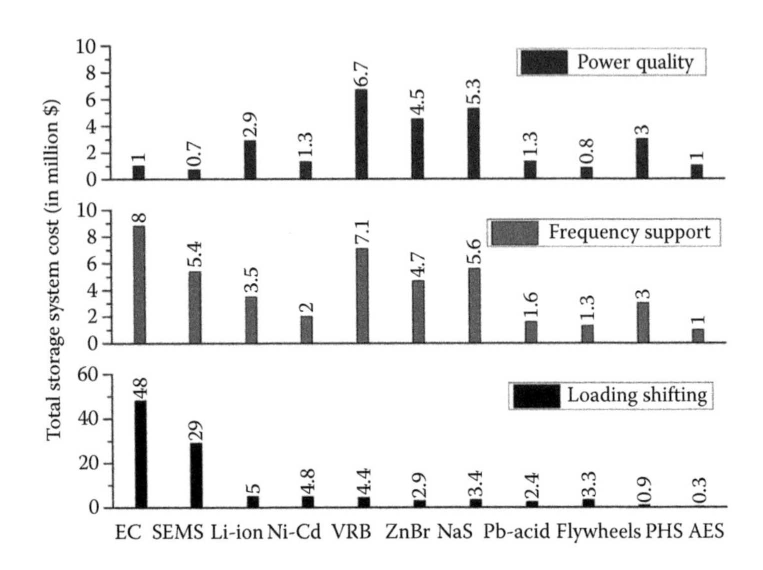

FIGURE 7.2 Total annual storage system cost for three different applications.

Li-ion and Ni-Cd batteries are economically viable for the power and frequency regulation applications. However, high material costs and the need for internal protection circuits increase their energy cost, which further increases their costs for load-shifting application. Although Pb-acid batteries have a lower cost and more lifetime, Li-ion batteries have several other advantages like high power-to-weight and high power-to-volume ratios that have led to their recent popularity.

SMES and ECs are the most attractive technologies for power quality application but are the least economically viable for load shifting application due to high material costs of superconductors and electrode. The large cost ratio of LIC is the cost of electrodes; therefore, more works should be focused on the synthesis cost of electrodes and develop novel electrodes with higher capacity and low cost (Figure 7.3).

7.2.3 Technical Evaluation

Aiming to the network model, the ESS should have three primary functions: leveling load, supplying a stable output power and controlling power quality. To keep the system operating continuously, the former two functions will be given sufficient consideration. Apart from that, in view of the technologies available and their suitability to power application, the energy response duration is a key issue to which more attention should be paid.

The energy storage capabilities of LICs are much greater than that of conventional capacitors, by about two orders of magnitude. The cycle life of a LIC is usually more than 500,000 cycles at 100% deep discharge, while its lifetime may exceed 12 years. Apart from this, a LIC can be easily charged and discharged in seconds. Furthermore, the energy efficiency is very high, ranging from 85% up to 98%. The

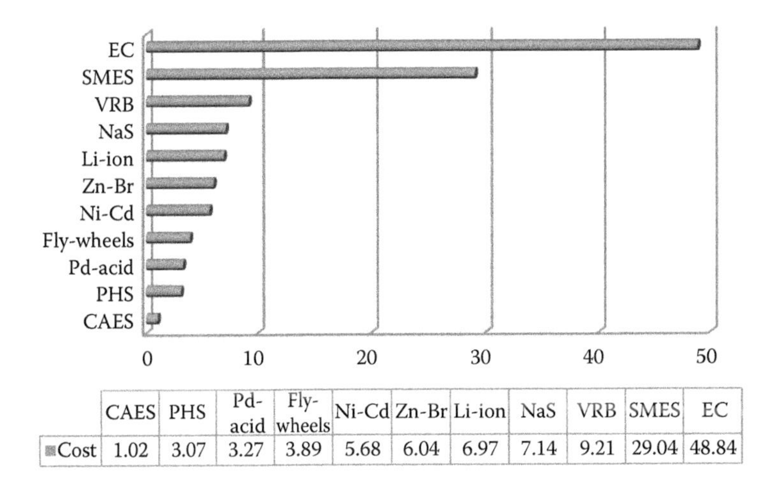

	CAES	PHS	Pd-acid	Fly-wheels	Ni-Cd	Zn-Br	Li-ion	NaS	VRB	SMES	EC
Cost	1.02	3.07	3.27	3.89	5.68	6.04	6.97	7.14	9.21	29.04	48.84

FIGURE 7.3 Total storage cost for combined application case.

power density of LICs is also high, reaching 10 kW/kg, which is a few orders of magnitude higher than that of batteries. However, the duration is very short due to the low energy density, and, it has a much higher self-discharge rate than batteries, which reaches 20% of the nominal energy per month.

Table 7.4 lists the main technical characteristics of the four ESSs [65]. Obviously, a small-scale SMES, FW, or SC can hardly supply enough electricity storage for hours to days. And that a large-scale SMES, FW, and SC are very expensive. Moreover, a RFB can achieve it easily. Hence, considering the balance between the energy density, power density, and function supplement, the three combined cases are technically feasible in a hybrid solar-wind system.

7.2.4 Evaluation of the Energy Storage Systems Cost

The cost calculation is under the condition that the maximum allowable imbalance power ratio is ±1%. The costs select the maximal admissible values and are neglecting the influence of the transformers and the insulation level of the system. The annual costs of the four ESSs are summarized in Table 7.2. From Table 7.2, the annual cost of the RFB is the lowest and that of the FW is the highest. And those of the SMES and SC are between that of the FW and the RFB.

From Figure 7.4, per unit costs have fallen quickly, with cost/kJ dropping faster than cost/Farad as voltage are increased. Maxwell Technologies claimed that in 2010, the cost of supercapacitors would have reached 0.4% per Farad, or $1.28 per kJ. With the rapid development of global LICs market, especially in China, for the low local producing cost, the cost of LICs will descend fast to an available level. IDTechEx believes that the progress of supercapacitors is much faster than that of lithium-ion batteries. By 2024, the global supercapacitor market value will reach 6 billion 500 million, and the market share is growing at the same time, thereby swallowing the battery market.

TABLE 7.4
Comparison of Technical Characteristics

	Discharge Time	Depth of Discharge (%)	Storage Duration	Energy Density (Wh/kg)	Power Density (W/kg)	Power Efficiency (P_{out}/P_{in}) (%)	Energy Efficiency (=E_{out}/E_{in}) (%)	Life Time (years)
SMES	Milliseconds–seconds	100	Minutes–hours	0.5–5	500–2,000	~90	>95	>20
FW	Milliseconds–minutes	100	Seconds–minutes	10–30	400–1,500	~80	>90	~15
SC	Milliseconds–minutes	100	Seconds–hours	2.5–15	500–5,000	~85	>95	>20
RFB	Seconds–hours	>75	Hours–months	10–30	16–33	~75	>70	5–10

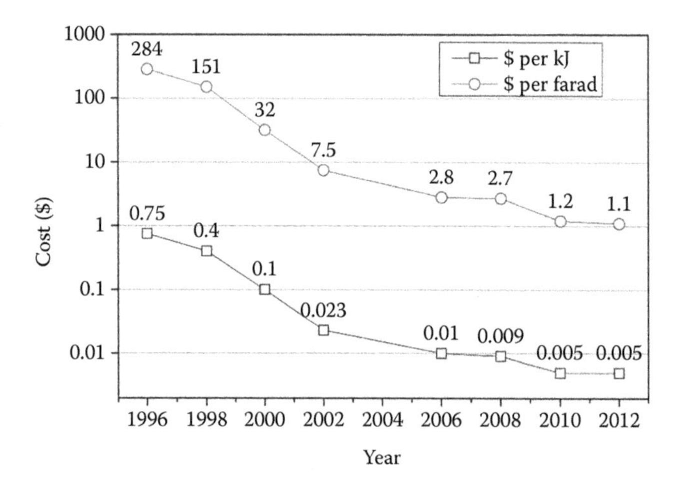

FIGURE 7.4 Change trend of supercapacitor costs in the past 15 years.

7.2.5 Business Challenge of Lithium-Ion Supercapacitors

Supercapacitors with superior performance, such as large capacity, high power, long life, low cost, and environment friendly, can be partial or total substitution of conventional chemical batteries, and have more widespread use than traditional chemical batteries. Technology of supercapacitors continues to develop; its applications range from the initial area extends to power electronic equipment and storage areas. In 2015, global ultracapacitors market reached $16 billion, it is forecasted that the future 5-year compound annual growth rate is expected to reach 21.3% (Figure 7.5).

In 2014, China market scale of supercapacitor (Figure 7.6) is 4.84 billion, 56.1% growth rate than 2013. In 2016, the UltraCapacitor market reached 8.3 billion. 2012–2015, the compound annual growth rate is more than 40%. Supercapacitors for transportation will be the most important driving force to support the rapid

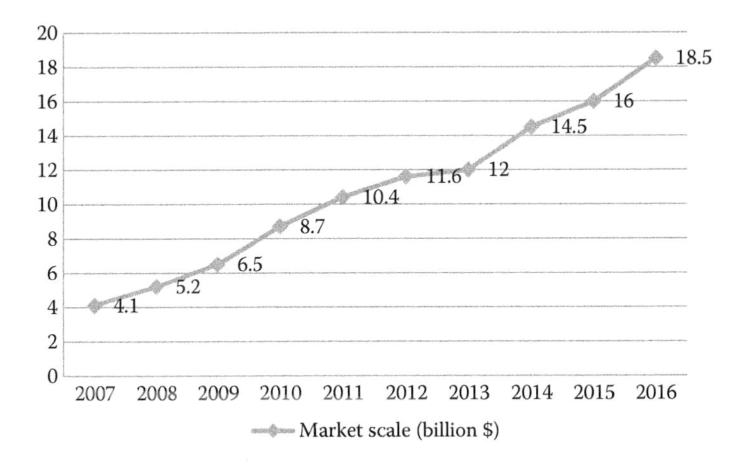

FIGURE 7.5 Global supercapacitor market scale from 2008 to 2016.

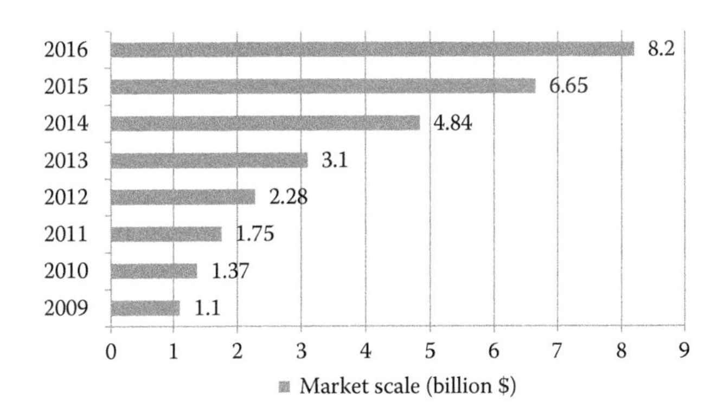

FIGURE 7.6 Supercapacitor market scale of China from 2009 to 2016.

development of the whole industry. In 2018, it will account for more than half of the total supercapacitor market share. Ultracapacitor industry market size in China in recent years is shown in the following figure:

During development for the next decade, supercapacitors will be an important part of the transportation industry and natural resources collection, in which supercapacitor used in a vehicle will become its main sales channels in the future (Figure 7.7).

7.2.6 Prospects of Global Lithium-Ion Capacitors Market

In 2015, Electro Standards Laboratories (ESL), Cranston, RI, a leader in the design of motor controls and power systems, has put their expertise in the energy field behind their offering of the ULTIMO Lithium-Ion Capacitor product line. The state of the art ULTIMO supercapacitors is the first commercialized LIC on the market and are quickly becoming the energy/power leader due to their superior energy density, charge retention, and higher cell voltage than traditional EDLC supercapacitors.

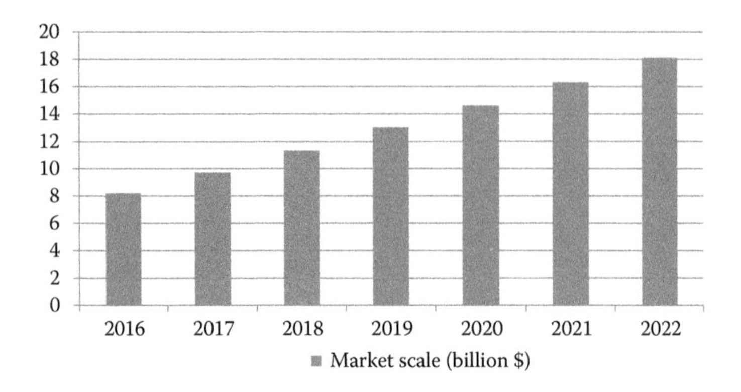

FIGURE 7.7 2016–2022 Ultracapacitor market size forecast in China.

These LIC cells also feature very high cycle life, fast charge/discharge times, high power density, long life, and low maintenance cost.

Electro Standards Laboratories has direct experience with pulsed power energy systems, load leveling for wireless electric vehicle charging, wave energy harvesting, and hybrid energy storage systems (HESS) that combine batteries, supercapacitors, and converter power electronics. The distribution activity of JSR Micro ultracapacitors by ESL has flourished due to growing demands in renewable energy, pulsed power applications, and hybrid energy storage modules (HESM) or HESS. The ULTIMO supercapacitor product line includes both laminate and prismatic individual cells as well as multi-cell modules. Both the laminate and prismatic designs differ greatly from the cylinder-shaped designs commonly found in the LIC market. The unique, flat designs of the ULTIMO Supercap cells allow for more compact stacking when used in series, making the most efficient use of space. Besides, ESL also offers an Energy Action System Lithium-Ion Capacitor Development Shelf, Model 3312, Cat# 331209 to compliment the ULTIMO product line. This shelf works with several LIC modules in series to provide connection of up to 9 LIC modules (108 LIC cells) along with associated protection, monitoring, and cell balancing circuitry. This unit can be incorporated into a larger system design or used by itself in applications requiring high power, high energy, and LIC energy storage. ESL can also supply custom power and module communication cables to accommodate each customer's individual application. They anticipated that its unique offering of ULTIMO supercapacitors, Action System development shelves, custom cabling, and engineering development services in battery, capacitor and hybrid systems will allow their customers to quickly realize the benefits of ULTIMO supercapacitors in their power systems/applications. ESL R&D services and its new ULTIMO product line provide its clients with a research and development partner that can assist in any phase of development from conception to production.

When using these anodes, limited by material properties, it is difficult to overcome the defects of material itself, such as graphite material, which has a poor compatibility with solvent and bad C-rate performance, hard carbon materials, which have low efficiency for the irreversible capacity in the first time is too big and not easy to process. These inherent defects themselves will greatly affect the performance of the

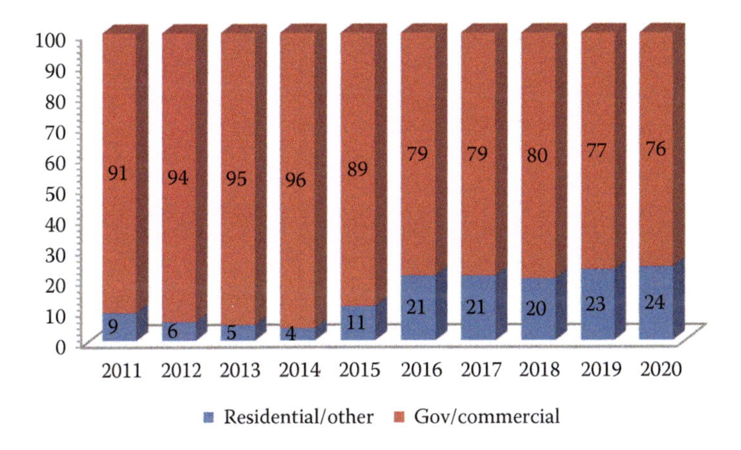

FIGURE 7.8 Lithium-ion capacitors (LICs) used in wind power generation global market value forecast (market share, %).

LIC unit. Therefore, to further improve the performance of the LIC, development of applicable anode materials is imperative.

At present, researches of LICs mainly focus on the production technology and electrode materials. The LIC has two kinds of parallel storage mode. To make the LIC further perfect and improved performance, a single electrolyte of EDLC or LIB is not adaptable. Therefore, we should develop the LIC electrolyte with the double function properties, to solve matching problem of anode/cathode materials and electrolytes.

Currently, these industries are mainly controlled by Japan and the United States, e.g., Japan Coke Li and Japan ACT company, respectively, master activated carbon and nano carbon technology; Japan Janebo company, Kureha chemical, and ATEC company master the anode techniques of polyacenes and hard carbon; UAS FERRO and German HONEYWELL occupy leading status in electrolyte, separator was controlled by Japan NKK monopoly, rich-porous current collector was controlled by three metal corporation monopoly.

Per ElectroniCast, the consumption value of LICs in wind power generation is forecast to reach over $150 million in 2021. ElectroniCast Consultants, a leading market and technology forecast consultancy, announced their market forecast of the worldwide consumption of LICs used in wind power (wind turbines) generation. ElectroniCast estimates that in 2011, the worldwide consumption value of LICs used in wind power generation applications was less than $1 million. From 2016 to 2021, the total consumption value forecasted an increase nearly 80% per year, reaching over $150 million in the year 2021 (Figure 7.8).

REFERENCES

1. Sundararagavan S, Baker E. Evaluating energy storage technologies for wind power integration. *Solar Energy*. 2012;86(9):2707–17.
2. Rheatsao. Maxwell Technologies delivers first commercial application of lithium-ion capacitor technology in China. EnergyTrend, a Business Division of TrendForce Corp. 2016.

3. Park B, Kim SJ, Sohn JS, Nam MS, Kang S, Jun SC. Surface plasmon enhancement of photoluminescence in photo-chemically synthesized graphene quantum dot and Au nanosphere. *Nano Research.* 2016;9(6):1866–75.
4. Wang R, Lang J, Zhang P, Lin Z, Yan X. Fast and large lithium storage in 3D porous VN nanowires–graphene composite as a superior anode toward high-performance hybrid supercapacitors. *Advanced Functional Materials.* 2015;25(15):2270–8.
5. Wang R, Lang J, Liu Y, Lin Z, Yan X. Ultra-small, size-controlled $Ni(OH)_2$ nanoparticles: Elucidating the relationship between particle size and electrochemical performance for advanced energy storage devices. *NPG Asia Materials.* 2015;7:e183.
6. Liu L, Shen B, Jiang D, Guo R, Kong L, Yan X. Watchband-like supercapacitors with body temperature inducible shape memory ability. *Advanced Energy Materials.* 2016;6(16):1600763.
7. Liu M-X, Gan L-H, Li Y, Zhu D-Z, Xu Z-J, Chen L-W. Synthesis and electrochemical performance of hierarchical porous carbons with 3D open-cell structure based on nanosilica-embedded emulsion-templated polymerization. *Chinese Chemical Letters.* 2014;25(6):897–901.
8. Xiao Y, Long C, Zheng M-T, Dong H-W, Lei B-F, Zhang H-R. et al. High-capacity porous carbons prepared by KOH activation of activated carbon for supercapacitors. *Chinese Chemical Letters.* 2014;25(6):865–8.
9. Meng Q, Wu H, Meng Y, Xie K, Wei Z, Guo Z. High-performance all-carbon yarn micro-supercapacitor for an integrated energy system. *Advanced Materials.* 2014;26(24):4100–6.
10. El-Kady MF, Ihns M, Li MP, Hwang JY, Mousavi MF, Chaney L. et al. Engineering three-dimensional hybrid supercapacitors and microsupercapacitors for high-performance integrated energy storage. *Proceedings of the National Academy of Sciences.* 2015;112(14):4233–8.
11. Xu X, Li S, Zhang H, Shen Y, Zakeeruddin SM, Graetzel M. et al. A power pack based on organometallic perovskite solar cell and supercapacitor. *ACS Nano.* 2015;9(2):1782–7.
12. Gao Z, Bumgardner C, Song NN, Zhang YY, Li JJ, Li XD. Cotton-textile-enabled flexible self-sustaining power packs via roll-to-roll fabrication. *Nature Communications.* 2016;7:12.
13. Wee G, Salim T, Lam YM, Mhaisalkar SG, Srinivasan M. Printable photo-supercapacitor using single-walled carbon nanotubes. *Energy & Environmental Science.* 2011;4(2):413–6.
14. Fu Y, Wu H, Ye S, Cai X, Yu X, Hou S. et al. Integrated power fiber for energy conversion and storage. *Energy & Environmental Science.* 2013;6(3):805–12.
15. Shi C, Dong H, Zhu R, Li H, Sun Y, Xu D. et al. An "all-in-one" mesh-typed integrated energy unit for both photoelectric conversion and energy storage in uniform electrochemical system. *Nano Energy.* 2015;13(Supplement C):670–8.
16. Ng CH, Lim HN, Hayase S, Harrison I, Pandikumar A, Huang NM. Potential active materials for photo-supercapacitor: A review. *Journal of Power Sources.* 2015;296(Supplement C):169–85.
17. Miyasaka T, Murakami TN. The photocapacitor: An efficient self-charging capacitor for direct storage of solar energy. *Applied Physics Letters.* 2004;85(17):3932–4.
18. Murakami TN, Kawashima N, Miyasaka T. A high-voltage dye-sensitized photocapacitor of a three-electrode system. *Chemical Communications.* 2005;26:3346–8.
19. Bae J, Park YJ, Lee M, Cha SN, Choi YJ, Lee CS. et al. Single-fiber-based hybridization of energy converters and storage units using graphene as electrodes. *Advanced Materials.* 2011;23(30):3446–9.
20. Zhang Z, Chen X, Chen P, Guan G, Qiu L, Lin H. et al. Integrated polymer solar cell and electrochemical supercapacitor in a flexible and stable fiber format. *Advanced Materials.* 2014;26(3):466–70.

21. Narayanan R, Kumar PN, Deepa M, Srivastava AK. Combining energy conversion and storage: A solar powered supercapacitor. *Electrochimica Acta.* 2015;178(Supplement C):113–26.
22. Xu J, Ku Z, Zhang Y, Chao D, Fan HJ. Integrated photo-supercapacitor based on PEDOT modified printable perovskite solar cell. *Advanced Materials Technologies.* 2016;1(5):1600074.
23. Liu W-W, Feng Y-Q, Yan X-B, Chen J-T, Xue Q-J. Supercapacitors: Superior micro-supercapacitors based on graphene quantum dots. *Advanced Functional Materials.* 2013;23(33):4111–22.
24. Wang K, Zhang X, Li C, Sun X, Meng Q, Ma Y. et al. Chemically crosslinked hydrogel film leads to integrated flexible supercapacitors with superior performance. *Advanced Materials.* 2015;27(45):7451–7.
25. Kramer M, Schwarz U, Kaskel S. Synthesis and properties of the metal-organic framework $Mo_3(BTC)_2$ (TUDMOF-1). *Journal of Materials Chemistry.* 2006;16(23):2245.
26. Chen H-W, Hsu C-Y, Chen J-G, Lee K-M, Wang C-C, Huang K-C. et al. Plastic dye-sensitized photo-supercapacitor using electrophoretic deposition and compression methods. *Journal of Power Sources.* 2010;195(18):6225–31.
27. Hsu C-Y, Chen H-W, Lee K-M, Hu C-W, Ho K-C. A dye-sensitized photo-supercapacitor based on PProDOT-Et2 thick films. *Journal of Power Sources.* 2010;195(18):6232–8.
28. Xu J, Wu H, Lu L, Leung S-F, Chen D, Chen X. et al. Integrated photo-supercapacitor based on Bi-polar TiO2 nanotube arrays with selective one-side plasma-assisted hydrogenation. *Advanced Functional Materials.* 2014;24(13):1840–6.
29. Murakami TN, Kawashima N, Miyasaka T. A high-voltage dye-sensitized photocapacitor of a three-electrode system. *Chemical Communications (Camb).* 2005(26): 3346–8.
30. Cohn AP, Erwin WR, Share K, Oakes L, Westover AS, Carter RE. et al. All silicon electrode photocapacitor for integrated energy storage and conversion. *Nano Letters.* 2015;15(4):2727–31.
31. Liu R, Liu Y, Zou H, Song T, Sun B. Integrated solar capacitors for energy conversion and storage. *Nano Research.* 2017;10(5):1545–59.
32. Cohn AP, Erwin WR, Share K, Oakes L, Westover AS, Carter RE. et al. All silicon electrode photocapacitor for integrated energy storage and conversion. *Nano Letters.* 2015;15(4):2727–31.
33. Sun Y, Yan X. Recent advances in dual-functional devices integrating solar cells and supercapacitors. *Solar RRL.* 2017;1(3–4):1700002.
34. Kazim S, Nazeeruddin MK, Grätzel M, Ahmad S. Perovskite as light harvester: A game changer in photovoltaics. *Angewandte Chemie.* 2014;53(11):2812–24.
35. Li C, Islam MM, Moore J, Sleppy J, Morrison C, Konstantinov K. et al. Wearable energy-smart ribbons for synchronous energy harvest and storage. *Nature Communications* 2016;7:13319.
36. Calió L, Kazim S, Grätzel M, Ahmad S. Hole-Transport materials for perovskite solar cells. *Angewandte Chemie International Edition.* 2016;55(47):14522–45.
37. Bagheri N, Aghaei A, Ghotbi MY, Marzbanrad E, Vlachopoulos N, Häggman L. et al. Combination of asymmetric supercapacitor utilizing activated carbon and nickel oxide with cobalt polypyridyl-based dye-sensitized solar cell. *Electrochimica Acta.* 2014;143:390–7.
38. Fujimoto M, Kida Y, Nohma T, Takahashi M, Nishio K, Saito T. Electrochemical behaviour of carbon electrodes in some electrolyte solutions. *Journal of Power Sources.* 1996;63(1):127–30.
39. Chung GC, Kim HJ, Yu SI, Jun SH, Choi JW, Kim MH. Origin of graphite exfoliation an investigation of the important role of solvent cointercalation. *Journal of the Electrochemical Society.* 2000;147(12):4391–8.

40. Zeng F, Ruan D, Fu G. Research progress of lithium ion capacitor industrial frontier technology. *Chemistry*. 2015;78(6):518–24.
41. Cao WJ, Zheng JP. Li-ion capacitors with carbon cathode and hard carbon/stabilized lithium metal powder anode electrodes. *Journal of Power Sources*. 2012;213(9):180–5.
42. Cao WJ, Zheng JP. Li-ion capacitors using carbon-carbon electrodes. *ECS Transactions*. 2013;45(29):165–72.
43. Kumagai S, Ishikawa T, Sawa N. Cycle performance of lithium-ion capacitors using graphite negative electrodes at different pre-lithiation levels. *Journal of Energy Storage*. 2015;2(8):1–7.
44. Ping LN, Zheng JM, Shi ZQ, Wang CY. Electrochemical performance of lithium ion capacitors using Li+-intercalated mesocarbon microbeads as the negative electrode. *Acta Physico-Chimica Sinica*. 2012;28(7):1733–8.
45. Zuo W, Li R, Zhou C, Li Y, Xia J, Liu J. Battery-supercapacitor hybrid devices: Recent progress and future prospects. *Advanced Science*. 2017;4(7):1600539.
46. Sivakkumar SR, Pandolfo AG. Evaluation of lithium-ion capacitors assembled with pre-lithiated graphite anode and activated carbon cathode. *Electrochimica Acta*. 2012;65(Supplement C):280–7.
47. Yuan M, Liu W, Zhu Y, Yongjin XU, Zhao F. Influence of Li intercalation mode on the performance of lithium ion capacitors. *Materials Review*. 2013;16.
48. Decaux C, Lota G, Raymundo-Piñero E, Frackowiak E, Béguin F. Electrochemical performance of a hybrid lithium-ion capacitor with a graphite anode preloaded from lithium bis(trifluoromethane)sulfonimide-based electrolyte. *Electrochimica Acta*. 2012;86(Supplement C):282–6.
49. Wang H, Yoshio M. Graphite, a suitable positive electrode material for high-energy electrochemical capacitors. *Electrochemistry Communications*. 2006;8(9):1481–6.
50. Wang H, Yoshio M. Performance of AC/graphite capacitors at high weight ratios of AC/graphite. *Journal of Power Sources*. 2008;177(2):681–4.
51. Sivakkumar SR, Nerkar JY, Pandolfo AG. Rate capability of graphite materials as negative electrodes in lithium-ion capacitors. *Electrochimica Acta*. 2010;55(9):3330–5.
52. Sivakkumar SR, Milev AS, Pandolfo AG. Effect of ball-milling on the rate and cycle-life performance of graphite as negative electrodes in lithium-ion capacitors. *Electrochimica Acta*. 2011;56(27):9700–6.
53. Schroeder M, Winter M, Passerini S, Balducci A. On the cycling stability of lithium-ion capacitors containing soft carbon as anodic material. *Journal of Power Sources*. 2013;238(238):388–94.
54. Karthikeyan K, Aravindan V, Lee SB, Jang IC, Lim HH, Park GJ. et al. A novel asymmetric hybrid supercapacitor based on Li_2 $FeSiO_4$ and activated carbon electrodes. *Journal of Alloys & Compounds*. 2010;504(1):224–7.
55. Zheng Z, Zhang P, Yan X. Progress in electrode materials for lithium ion hybrid supercapacitors. *Chinese Science Bulletin*. 2013;58(31):3115–23.
56. Lee JH, Shin WH, Lim SY, Kim BG, Choi JW. Modified graphite and graphene electrodes for high-performance lithium ion hybrid capacitors. *Materials for Renewable and Sustainable Energy*. 2014;3(1):22.
57. Zhang F, Zhang T, Yang X, Zhang L, Leng K, Huang Y. et al. A high-performance supercapacitor-battery hybrid energy storage device based on graphene-enhanced electrode materials with ultrahigh energy density. *Energy & Environmental Science*. 2013;6(5):1623–32.
58. Kim HU, Sun YK, Shin KH, Jin CS. Synthesis of $Li_4Mn_5O_{12}$ and its application to the non-aqueous hybrid capacitor. *Physica Scripta*. 2010;139(9):561–78.
59. Jang BZ, Liu C, Neff D, Yu Z, Wang MC, Xiong W. et al. Graphene surface-enabled lithium ion-exchanging cells: Next-generation high-power energy storage devices. *Nano Letters*. 2011;11(9):3785.

60. Han P, Ma W, Pang S, Kong Q, Yao J, Bi C. et al. Graphene decorated with molybdenum dioxide nanoparticles for use in high energy lithium ion capacitors with an organic electrolyte. *Journal of Materials Chemistry A*. 2013;1(19):5949–54.
61. Hu X, Deng Z, Suo J, Pan Z. A high rate, high capacity and long life ($LiMn_2O_4$ + AC)/$Li_4Ti_5O_{12}$ hybrid battery–supercapacitor. *Journal of Power Sources*. 2009;187(2):635–9.
62. Hao YJ, Wang L, Lai QY. Preparation and electrochemical performance of nano-structured $Li_2Mn_4O_9$ for supercapacitor. *Journal of Solid State Electrochemistry*. 2011;15(9):1901–7.
63. Tang W, Hou Y, Wang F, Liu L, Wu Y, Zhu K. $LiMn_2O_4$ nanotube as cathode material of second-level charge capability for aqueous rechargeable batteries. *Nano Letters*. 2013;13(5):2036–40.
64. Yang PX, Cui WY, Li LB, Liu L, An MZ. Characterization and properties of ternary P(VdF-HFP)-LiTFSI-EMITFSI ionic liquid polymer electrolytes. *Solid State Sciences*. 2012;14(5):598–606.
65. Ren L, Tang Y, Shi J, Dou J, Zhou S, Jin T. Techno-economic evaluation of hybrid energy storage technologies for a solar–wind generation system. *Physica C Superconductivity & Its Applications*. 2013;484(1):272–5.

Index

Page numbers in italic indicate a figure and page numbers in bold indicate a table on the corresponding page

F

G

H

U

V

W

X

Z